Special Relativity for Beginners

A Textbook for Undergraduates

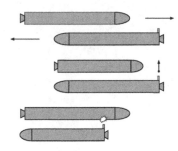

Special Relativity for Beginners

A Textbook for Undergraduates

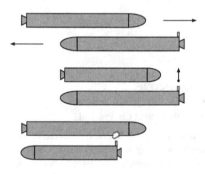

Jürgen Freund

Schubart College, Germany

 **World Scientific**

NEW JERSEY · LONDON · SINGAPORE · BEIJING · SHANGHAI · HONG KONG · TAIPEI · CHENNAI

Published by

World Scientific Publishing Co. Pte. Ltd.

5 Toh Tuck Link, Singapore 596224

USA office: 27 Warren Street, Suite 401-402, Hackensack, NJ 07601

UK office: 57 Shelton Street, Covent Garden, London WC2H 9HE

British Library Cataloguing-in-Publication Data
A catalogue record for this book is available from the British Library.

SPECIAL RELATIVITY FOR BEGINNERS
A Textbook for Undergraduates

Copyright © 2008 by World Scientific Publishing Co. Pte. Ltd.

ISBN-13 978-981-277-159-9
ISBN-10 981-277-159-X
ISBN-13 978-981-277-160-5 (pbk)
ISBN-10 981-277-160-3 (pbk)

Typeset by Stallion Press
Email: enquiries@stallionpress.com

Printed in Singapore.

Foreword

An oft-cited anecdote reports how Sir Arthur Eddington was admired by one of his assistants for being one of only three men on earth to thoroughly understand Albert Einstein's theory of relativity.

In 1987, when the special theory of relativity was already more than eighty years old, his former colleague and biographer, Abraham Pais, updated this number in an address on occasion of the 900th anniversary of the University of Bologna saying, "At an optimistic estimate, only one in one hundred thousand people living today understands the essence of Einstein's theory of relativity."

The special theory of relativity, the exclusive subject of this book, compares the experiences of observers in rectilinear and uniform motions relative to each other. It is based on two principles:

The first one, the so-called principle of relativity, requires that for two observers in rectilinear and uniform motion *all* laws of physics have to assume the same form. In a pictorial representation: Inside a railroad car in rectilinear and uniform motion with windows shaded to obscure an outward view, *no* physical experiment, neither the oscillation of a pendulum, nor the boiling of water, nor the ticking of a Geiger counter, shall permit a measurement of the car's velocity relative to the railway track.

The second principle purports that each observer measures the same speed of light, c, whether the light is emitted from a body at rest or from a body in rectilinear and uniform motion. While the principle of relativity is "only" a generalization of the principle of relativity of Galilean–Newtonian mechanics to the whole of physics, the principle of the invariance of the speed of light places the true demand on our imagination. While common sense tells us that a passenger walking forward at 5 kph in a train passing us at 130 kph proceeds at 135 kph relative to us, this is not true for light! When a passenger shines a torch, the light propagates with a speed of c, both relative to the train and relative to the railway track, as evidenced by experiment.

Since the principle of relativity poses a kinematic demand on all laws of nature, the special theory of relativity is not merely a theory alongside mechanics, electrodynamics etc, but all branches of physics have to be examined for their ability to satisfy both fundamental principles of special relativity. Maxwell's equations of electrodynamics do fulfill these requirements, but classical mechanics does not. Its adaptation to the principles of relativity leads to many consequences of which the famous formula $E = mc^2$ is the best-known example. Even Pauli's exclusion principle – to mention just one out of many examples from quantum mechanics – describing the electron shell structure and thus laying the foundation to, strictly speaking, all of chemistry is deeply rooted in the special theory of relativity.

Minkowski's formulation of the theory of relativity within the bounds of a four-dimensional space-time geometry is not only known as its mathematical perfection. It also entails a considerable simplification of our worldview by opening our eyes to the four-dimensional unity not only of space and time but also of momentum and energy, force and power, electric current density and charge density etc. Finally, it allows a very plastic depiction of space-time dependencies in the form of light cones.

The focus today is by no means an experimental verification of the special theory of relativity. From its very beginning, its "intrinsic perfection" has been such a strong argument for Einstein that, when putative contradictions between theory and experiment arose, he preferred to doubt the experiment rather than his theory. The theory has stood all "external tests", and it still stands them in current experiments with increasing precision. With the construction of accelerators or the synchronization of clocks in the global positioning system, the special theory of relativity has turned into an engineering science, to mention just two examples. This theory, which marked the beginning of a development commonly called "modern physics", has thus become classical by itself.

As a science, special relativity has never been vexed with paradoxes, but as a subject of instruction it has. Paradoxes like the twin paradox exemplify that comprehension and teachability of a theory requires time, all the more so because notions as familiar as space and time are shaken to their foundations. For didactical reasons these paradoxes are still welcome, even though they have long ceased to be paradoxes, because students are thus trained to profoundly ponder on the theory. This also points to the main difficulty encountered by learners, to wit transcending his or her habits of thought. Moreover, every teacher knows from experience how carefully language has to be handled in order to explain a theory without having a premature recourse to the formalism used therein. In the case of special relativity, mathematics is not the chief problem of learners, after all. According to Sexl, "A rudimentary knowledge of root extraction suffices to comprehend the mathematical derivations."

In German universities special relativity is, of course, part of the regular curriculum. Before his or her diploma exam, every student is exposed to its foundations in the mechanics or electrodynamics course. In most cases, however, this is not sufficient for developing a "physical intuition" as to why effects such as time dilation and length contraction actually appear. Students searching for deeper insights are referred to special courses.

Special relativity has for a long time entered the curricula of secondary schools as well, although only to an extent that cannot do justice to the importance of the theory and to the overwhelming interest of students. We can understand these restrictions without endorsing them. Still today, many a physics student can embark on a high school teaching career without sufficient acquaintance with the theory of relativity, or even its didactics. This is also true for older colleagues. Some school board officials still try to solve the constraints imposed by the reciprocity of lessons needed and lessons given by omitting relativity altogether, arguing that with its thought experiments and Minkowski diagrams relativity is not physics proper. Such "arguments" can easily be invalidated. Thought experiments and the connection between physics and geometry are typical methodical elements, not exclusively but mostly, of 20th century physics, suitable for a rather profound demonstration in high school, as exemplified by special relativity. Furthermore, this very theory with its operational prescriptions like clock synchronization strongly focuses our attention on measurability and the difference between measurement and observation.

This status quo description of relativity in German high schools manifests both a need to improve the university education of students enrolled in teacher programs and to enhance the in-service training of teachers on active duty. And every thoughtful didactical approach not only adopts innovations but mediates between different cognition and speech levels.

The author of this textbook entitled "relativity for beginners" lives up to these demands. He addresses, above all, first year physics students who, even before their regular courses in theoretical physics, want to delve into special relativity. For students in teacher programs this book contains all subject matters needed to teach and all additional background knowledge. Chapters 1 to 4, 7 to 9, 15, 20, 23, 24, 26, and 27, as well as parts of Chapters 6, 14, and 29 offer a good basis for experienced teachers trying to prepare a course of at least 15 hours duration. And this book is even detailed enough for gifted high school juniors and seniors to browse through profitably. With the exception of Part IV, they can even follow the mathematical derivations line by line which are not too difficult but at times longer than what they are used to.

This book deserves widespread acceptance in both universities and high schools thus giving the author the readership envisaged.

Karl-Heinz Lotze
(Professor for Physics and Astronomy Didactics)
Jena, April 2004

Preface

Since its beginning one hundred years ago, the special theory of relativity has become a cornerstone of modern physics. Over the course of time it has been scrutinized in a multitude of experiments and has always been verified with high accuracy. The correctness of this theory can no longer be called into question.

Right after its discovery by Albert Einstein in 1905, special relativity was only gradually accepted because it made numerous predictions contradicting common sense, fervently castigated by Einstein, and also defied experiment for too long a time. It was only with the advent of particle or high energy physics that matter could be accelerated to very high velocities, close to the speed of light, which not only verified special relativity but also made it a requirement for machine construction. The expansion of the physical research frontier toward astronomy and cosmology during the past ten to twenty years considerably increased the importance of special relativity and, above all, general relativity based thereupon. Since astrophysics has in the same time become very popular among readers with a scientific background, the two theories of relativity have attained unprecedented publicity. The fascination with astronomy of children and youths shall only be mentioned incidentally, it is, however, one of the most impressive features of schools today.

And still, the special theory of relativity is only of marginal importance in both high school and university curricula – not to mention general relativity. Both have an aura of incomprehensibility, thus being reserved for specialists only. The transfiguration of Einstein's person into the greatest genius in science history hardly contributes to lowering the mental hurdles against his achievement. Therefore, a popularized, yet substantial and comprehensive treatise on special relativity, written for readers with a background in mathematics and physics, is needed.

This textbook for beginners attempts to secure easy access to special relativity. It addresses all students who have passed with ease their first undergraduate physics courses and enjoyed them. Even high school graduates, who excelled in mathematics and physics, may study this book unless they shy away from making up for unfamiliar mathematical formalism with the help of an additional textbook.

It must then be a discovery for readers that the notorious intricacies of special relativity are not rooted in mathematics. Unfamiliar, on the contrary, is the rigorous thinking at times required before a very simple formula can be committed to paper or interpreted. For the problems of relativity cannot be solved as mechanically as problems of pre-relativistic physics. Lest the red thread be lost, a textbook of relativity for beginners has to ponder hither and thither when our imagination leads us astray, when we grasp for orientation in the thicket of inertial frames, contractions, and dilations.

It is not below the standard of a textbook for beginners to expound problems detailed enough for readers to follow line by line. Desultory calculations are often considered an affront by those beginners who are apt to study everything down to the last detail, i.e. by the very best. They would prefer to see new and unfamiliar matter being presented by a few sample calculations. Studying physics is, in a sense, comparable to learning foreign languages. But no language textbook is limited to simply unfolding grammar and vocabulary, leaving the first construction of sentences to the learner's judgement. Therefore, the reader is urged to duplicate the calculations, if necessary, with paper and pencil, should this treatment not be exhaustive enough.

A first glance at this book should not deter anyone from grappling with the concepts disseminated herein, just because in some chapters the formulas seem to exceed the text lines. The more formulas appear in a scientific text the easier it is to grasp! Nothing is worse for beginners than derivations shortened by well-intentioned sentences. Such a procedure turns science into a secret science for the initiated. Formulas are, in the true sense of the word, democratic and international! Only for popular science texts, written for quite a different readership, is the famous bon mot applicable that with every additional formula the number of prints sold is halved. With approximately 1000 formulas in this text the potential readership would have to be astronomically large for just one reader to be left after a reduction by a factor of 2^{1000}. But maybe there is also a reciprocal law saying that the number of prints sold doubles with each figure. Then those 100 figures in the text would not only be of didactical value but also promote the sales of this book!

To aid readers without any previous knowledge of relativity, all chapters containing only additional information that can be skipped are marked by an asterisk. This reduces the text by one half. For students preparing for exams as well as for scientists in need of a formulary on special relativity the gray shaded areas are of assistance in finding their way through this book.

The present textbook on special relativity comes to the point directly. The historical development of the late nineteenth century leading to the special theory of relativity is certainly intriguing but its knowledge is by no means a prerequisite for understanding the theory. Even Einstein reportedly ignored it. Readers appreciating this cultural-historical approach are given ample opportunity to become acquainted with it in many books. Instead, such concepts as time dilation, length contraction, Lorentz transformation, and Minkowski diagrams are introduced right at the beginning and in great detail on the level of modern high school texts. After the explanation of simultaneity and the addition theorem of velocity this book leaves, apart from minor exceptions, the scope of advanced secondary school books and turns to the more advanced topics, all the way to the four-vector formalism and the principles of invariance. The pace is thus increased and adjusted to the reader's growing ability to cope with the mathematics involved. The full tensor formalism, however, is dispensed with because it is not really needed before general relativity is introduced. Establishing tensor calculus prematurely is apt to deter readers altogether and is certainly one of the chief reasons why relativity suffers from an elitist smack.

The attentive reader cannot fail to notice that some crucial results of special relativity are derived more than once using varying approaches, e.g. the Lorentz transformation in one spatial direction is derived three times in Chapters 6 and 15, the Doppler formula four times (Chapters 14, 15, and 31), the Lorentz transformation in two directions twice in Chapter 18, also twice the unification of electric and magnetic forces in Chapters 5 and 33, the velocity addition formula (Chapters 9 and 30), as well as the aberration formula (Chapters 10 and 31). This may be called a redundancy or a gain. Beginners will certainly be glad to find many routes to the goal, as is so often in physics. But for relativity, having had to fight so long for acceptance, this is of fundamental importance; it is self-contained and without contradictions.

It is to be hoped that many students will improve their ability in the mathematical methods of physics by working through this book. Beyond that they will not probably learn much physics of use for their theses and subsequent activities. But that is not the point here. On the contrary, the reader is to be carried away into a world that is, on superficial inspection, outlandish, paradoxical, even outright absurd. The intellectual penetration of these supposed absurdities makes up the appeal of the theory of relativity, it widens the horizon, it forms a new world view in the classical sense of the word. In this sense I wish you enjoy it!

Finally, I would like to express my gratitude to Emil Wiedemann from Fürth who took time to thoroughly and critically read my manuscript in all stages. His comments inspired me time and again.

Jürgen Freund
Aalen near Ulm, 14th March 2004
(Einstein's 125th Anniversary)
Contact: freund@relativitaet.info

P. S. (2nd January 2006) My special thanks go to Bill Beck from Seattle for his continued assistance in finding the most professional translations from German to English.

P. P. S. (4th January 2008) I want to express my sincere gratitude to Professor Kok Khoo Phua, Chairman of World Scientific, for his personal intervention on behalf of my manuscript, and to Alvin Chong, desk editor, for his admirable patience in view of my perfectionism.

Contents

PART I
Introduction

Chapter 1

The Postulates of the Special Theory of Relativity

Imagine a railroad station with six tracks (Fig. 1.1): On track **1a** a train has stopped, the train on track **1b** is going to the east at a velocity of 20 km/h. On track **2a** a train is going to the west at 20 km/h, while the train on track **2b** is at rest. On tracks **3a** and **3b**, trains are underway at 10 km/h to the west and to the east, respectively. All velocities are measured relative to the tracks; negative signs are assigned to westbound velocities.

Between tracks **1b** and **2a**, as well as between tracks **2b** and **3a**, distances are long. Therefore, travelers can always see one neighboring train only, but they cannot see the tracks. The observations made are indistinguishable for all travelers, i.e. travelers in trains on the a-tracks see b-trains go to the east at a velocity of 20 km/h, travelers in b-trains see a-trains go to

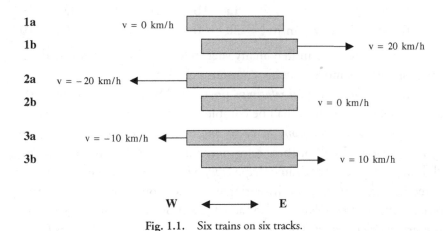

Fig. 1.1. Six trains on six tracks.

the west at 20 km/h. The movements on the three pairs of tracks can only be distinguished if passengers can observe the tracks. Thus, the tracks constitute an *absolute reference frame*. The existence of such an absolute reference frame is a prerequisite for distinguishing these three processes which have in common that the relative velocity between each two trains is the same (20 km/h).

Now let us imagine the trains are being replaced by spaceships in empty space. There are no longer any material objects, such as tracks or stars, which can be used as absolute reference frames. Evidently, all three patterns of motion become indistinguishable again. In the absence of an absolute reference frame, an *absolute velocity* (such as 0 km/h, 10 km/h to the west, etc.) can no longer be assigned to an object. There are only *relative velocities* between the spaceships, and in all three cases they are 20 km/h to the east or 20 km/h to the west, respectively.

Now all six spaceships shall be mutually visible. We now have the choice to *declare any spaceship a reference frame* in which we can measure the velocities of all other spaceships relative to the spaceship chosen. Choosing, for instance, spaceship **1a** as the reference frame, referred to as *1a-frame*, the six spaceships considered above have these relative velocities:

Spaceship	1a	1b	2a	2b	3a	3b
Relative Velocity (km/h)	0	20	−20	0	−10	10

In the *1b-frame*, the spaceships have these relative velocities:

Spaceship	1a	1b	2a	2b	3a	3b
Relative Velocity (km/h)	−20	0	−40	−20	−30	−10

Of course, there are infinitely many such reference frames associated with other spaceships moving at any constant velocity.

The *principle of special relativity*, postulated by Einstein, purports that all such reference frames shall be completely equivalent, i.e. *no such frame shall be distinguishable from, or be in any way superior to, any other frame by any property*. In all of these reference frames one shall be able to write all laws of physics in the same shape (*shape invariance*). Since their motion is rectilinear, uniform, and irrotational (i.e. without any acceleration), all objects left to themselves, i.e. not acted upon by any forces, are subject to the *principle of inertia*. They are therefore called *inertial (reference) frames*.

Principle of special relativity: All reference frames in rectilinear, uniform, and irrotational motion, i.e. all so-called inertial (reference) frames, shall be completely equivalent in physics. No inertial frame shall be distinguished from any other inertial frame by any property.

The choice of the verb "shall" expresses that this property is being postulated at this point. Only experience, i.e. experiment, will prove or disprove the correctness of this demand.

Everything said so far does not sound very exciting because it does not clash with *common sense*. Einstein's second postulate, however, *the principle of the invariance of the speed of light*, is quite different. To understand this postulate better, let us make a short detour into acoustics.

Sound propagates in a medium such as air at the *speed of sound*, v_s. If the source of sound is at rest with respect to the medium of propagation, the speed of sound is *isotropic*, i.e. the same in all directions. If the source of sound moves at the speed v relative to the air, e.g. to the east, the sonic speed v_s' relative to the source of sound is:

$$v_s' = v_s + v \quad \text{for propagation to the west,}$$
$$v_s' = v_s - v \quad \text{for propagation to the east.}$$

Quite obviously, the propagation of sound is no longer isotropic with respect to the source of sound (Fig. 1.2).

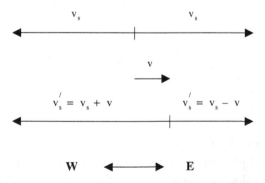

Fig. 1.2. The speed of sound with respect to the source of sound is changed when the source of sound moves to the east at a speed of v.

In the nineteenth century, light was believed to propagate in a similar way in the medium called *ether*. The ether was postulated to *permeate the entire universe*, even massive material objects. Thus, the hypothetical ether became an absolute reference frame like the tracks in our previous example. It was assumed as a matter of fact – being compatible with *common sense* – that the speed of light depends on the motion of the light source relative to the ether. Therefore, the search for the ether and an experimental verification of the *anisotropy* of light propagation became pivotal questions of nineteenth century experimental physics.

In 1887, *Michelson and Morley* devised an experiment that became crucial for the development of physics. Let us make the following consideration (Fig. 1.3):

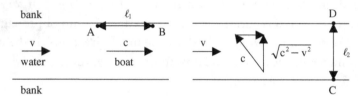

Fig. 1.3.   A boat sailing on a river: downstream (left) and at right angles to its banks (right).

A river flows from left to right at a velocity of v relative to the banks. A boat on the river is sailing with the velocity c relative to the water. When the boat is sailing downstream (or upstream), its velocity relative to the banks is $c + v$ (or $c - v$). To cover the distance ℓ_1 from A to B and back, the time

$$t_1 = \frac{\ell_1}{c + v} + \frac{\ell_1}{c - v} = \frac{2\ell_1 c}{c^2 - v^2} = \frac{2\ell_1}{c} \frac{1}{1 - \dfrac{v^2}{c^2}} \tag{1.1}$$

is required. When the boat is sailing at right angles to the banks, it has to turn its bow slightly upstream and thus reaches the velocity $\sqrt{c^2 - v^2}$ relative to the banks. For the distance ℓ_2 from C to D and back, the boat requires

$$t_2 = \frac{2\ell_2}{\sqrt{c^2 - v^2}} = \frac{2\ell_2}{c} \frac{1}{\sqrt{1 - \dfrac{v^2}{c^2}}}. \tag{1.2}$$

If v is small compared with c, *Taylor expansions* of the velocity factors of (1.1) and (1.2) can be made to give:

$$\left(1 - \frac{v^2}{c^2}\right)^{-1} \approx 1 + \frac{v^2}{c^2} + \cdots$$

$$\left(1 - \frac{v^2}{c^2}\right)^{-1/2} \approx 1 + \frac{1}{2}\frac{v^2}{c^2} + \cdots$$

This gives a running time difference of

$$\Delta t = t_1 - t_2 \approx \frac{2\ell_1}{c}\left(1 + \frac{v^2}{c^2}\right) - \frac{2\ell_2}{c}\left(1 + \frac{1}{2}\frac{v^2}{c^2}\right). \tag{1.3}$$

Now let us imagine the water is replaced by the ether moving at the velocity v from left to right through an interferometer at rest on earth. The boat is replaced by a light ray moving at the *speed of light*, c, relative to the ether. The light ray is cast on a semitransparent mirror S and split into two rays. One ray returns to S via mirror M_1, the other one moves at right angles to the first and returns to S via mirror M_2. After passing S, both rays run parallel and interfere until they reach the observer O (Fig. 1.4).

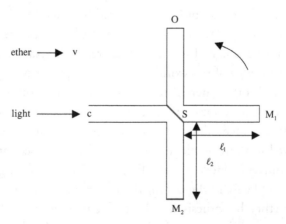

Fig. 1.4.   Michelson interferometer.

When this so-called *Michelson interferometer* is turned by 90°, the two rays are interchanged and the new running time difference becomes

$$\Delta t' = t_1' - t_2' \approx \frac{2\ell_1}{c}\left(1 + \frac{1}{2}\frac{v^2}{c^2}\right) - \frac{2\ell_2}{c}\left(1 + \frac{v^2}{c^2}\right). \qquad (1.4)$$

So the two running time differences differ by

$$\Delta t - \Delta t' = \frac{\ell_1 + \ell_2}{c}\frac{v^2}{c^2}. \qquad (1.5)$$

On turning the interferometer, *the interference pattern observed at O should alter. But no change occurs!* One might argue that the measurement is being made when the earth is at rest relative to the ether, i.e. for $v = 0$. But due to the earth's annual orbit around the sun, earth would move at 60 km/s relative to the ether six months later, and with $\ell_1 = \ell_2 = 10\,\mathrm{m}$ for the lengths of the interferometer arms a value of $2.7 \times 10^{-15}\,\mathrm{s}$ would be obtained for (1.5). Since light with a wavelength of $\lambda = 600\,\mathrm{nm}$ has a period of about $2.0 \times 10^{-15}\,\mathrm{s}$, a change of the interference pattern should be observable. But again, *no effect can be observed.*

The result of the experiment was that *at no time could a motion of the earth relative to the hypothetical ether be proven.* Any light signal emitted from any light source on earth propagates in all directions, i.e. isotropically, with the same speed, $c = 299{,}792{,}458\,\mathrm{m/s}$. In subsequent decades, the Michelson–Morley experiment was repeated with increasing accuracy, but the ether hypothesis could not be verified and has therefore been rejected. Since no special role in the universe can be attributed to the earth, since the earth cannot be assigned to a preferred reference frame according to our present knowledge of the world, the *isotropy of light propagation* must also be true for all other inertial reference frames. It should be observed that the earth is, strictly speaking, not an inertial reference frame because it rotates on its axis once a day and revolves around the sun once a year and is even involved in more accelerated motions in our galaxy and with our galaxy in the universe. However, the fictitious forces resulting from these motions are completely negligible for the experiment described.

After the ether hypothesis had been falsified, *Ritz* in 1908 continued the discussion with the convincing proposition that light propagates isotropically from every light source but reaches light receivers with different velocities depending on the relative velocity between the source and the receiver. Light from a star moving away from earth should reach earth with a lower velocity than light from a star approaching earth.

This hypothesis about light propagation could also be falsified. Light from a *binary star* – two stars revolving around a common center: one towards the earth, the other one away – should reach the earth with different velocities. Under certain circumstances one such star should even be observable twice at the same time: moving toward the earth (higher speed of light) and having moved away from the earth some time earlier (lower speed of light). However, such observations have never been reported.

Yet another unmistakable hint on the inaccuracy of Ritz's hypothesis is recent. In 1987, a *supernova explosion* happened in the large Magellanic cloud 165,000 light-years away, and *neutrinos* were emitted by nuclei. Neutrinos, like photons, travel at the speed of light. Although the nuclei's velocities differed by as much as 10,000 km/s, all neutrinos arrived within a time interval of 10 s. So there is no alternative to *Einstein's second postulate*:

> Principle of the constancy[a] of the speed of light: The speed of light in a vacuum has the same value in each inertial reference frame, irrespective of the velocities of the light source or the light receiver. It is a fundamental physical constant amounting to $c = 299,792,458$ m/s.

The magnitude of c was not known to Einstein to such an accuracy, as a matter of fact.

In concrete terms, the second principle has the following meaning: A star moves away from earth at half the speed of light. The light emitted by that star toward earth does *not* leave the star at one and a half times the speed of light (ether hypothesis), but at the speed of light. It arrives on earth at the speed of light, *not* at one half the speed of light (Ritz's hypothesis). A different star approaches earth at half the speed of light. The light emitted by that star also leaves it at the speed of light and also reaches earth at the speed of light.

---

[a] Strictly speaking, we should talk about the *invariance* of the speed of light. However, we follow common usage by calling it *constancy*.

This principle is in obvious violation of all common sense! Einstein's ingenuity is splendidly manifested by the fact that he made this very principle a keystone of his great theory which, at the time of its discovery in 1905, was by no means verifiable by experiment. The decade-long, at times rather fierce, rejection of his theory of relativity is, to some extent, rooted in the apparent preposterousness of this principle and the conclusions derived from it.

One of the reasons for the overwhelming fascination emanating from the special theory of relativity, once fully understood, is the fact that *only two principles underlie it*, the principle of special relativity and the principle of the constancy of the speed of light.

With three exceptions, the *invariance of electric charge* (Chapter 5), the *invariance of transverse momentum* (Chapter 20), and the formulation of *forces in electric and magnetic fields* (Chapter 34), no further working hypotheses are proposed in the course of this book, and no experimental findings are employed for the construction of this theory. Consequently, the procedure is of a *purely deductive nature*: All phenomena are predicted by theory and, if necessary, subsequently confirmed by independent experiment.

Chapter 2

Time Dilation

From now on, we will often make use of *two coordinate systems*: The first, called the S-frame or S-system, is a Cartesian coordinate system with *three orthogonal spatial coordinate axes* (x, y and z) and one *time coordinate*, t. The other one, called the S'-frame or S'-system, also has three orthogonal spatial coordinate axes (x', y' and z'), *parallel* to the axes of the S-frame, and a time coordinate, t' (Fig. 2.1). S' moves at the *constant velocity* **v** relative to S. For most problems in this book $\mathbf{v} = (v, 0, 0)$, i.e. a *one-dimensional motion*, is sufficiently general. For simplicity, we take $t = t' = 0$ when the origins of both the unprimed frame and the primed frame coincide.

Let us now embark on a thought experiment with a *light clock*. A light clock (Fig. 2.2) is a box 0.15 m tall, at both ends equipped with perfect

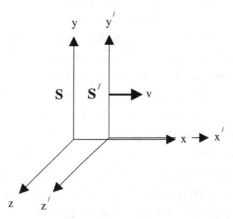

Fig. 2.1.   The S-frame and the S'-frame.

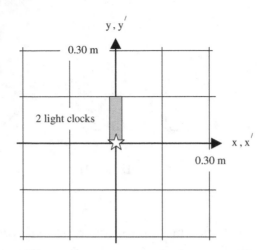

Fig. 2.2.   The light flash and the light clocks at $t = 0$ s.

mirrors. At the lower end, a light flash is being ignited which gets reflected very many times between the two mirrors. For one round-trip the light requires one nanosecond (ns $= 10^{-9}$ s). Thus, each time the reflected light returns to the lower mirror, the display of a counting mechanism calibrated in nanoseconds proceeds by one unit.

The origins of both coordinate frames are equipped with one light clock each, stretching from $y_1 = y_1' = 0.00$ m to $y_2 = y_2' = 0.15$ m. An unimpaired mutual penetration of the light clocks is presupposed, of course. When both origins coincide, i.e. at $t = t' = 0$, a light flash is being ignited at the origins, henceforth beating time in *both* light clocks.

After 0.5 ns, the light flash has turned into a *spherical wave* with a radius of 0.15 m. When we give the S'-frame a speed of $v = c/\sqrt{2}$, the origin of S' has moved about 0.11 m to the right. The spherical wave is now arriving at the upper mirror of the light clock in the S-frame. According to the above assumptions, a time of $t = 0.5$ ns has elapsed there. But the spherical wave has not yet reached the upper mirror of the light clock in the S'-frame. Therefore, $t' < 0.5$ ns must be true in that frame (Fig. 2.3). It seems that time in the S'-frame passes more slowly than time in the S-frame.

After $t = 0.5\sqrt{2}$ ns, the radius of the spherical wave has grown to about 0.21 m. The origin of S' has moved 0.15 m to the right. The spherical wave

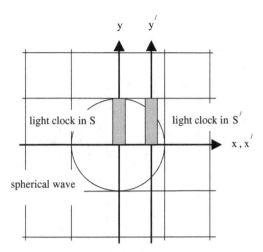

Fig. 2.3. The light clocks at $t = 0.5\,\text{ns}$.

is now touching the upper mirror of the light clock in the S'-frame, so that the time $t' = 0.5$ ns has elapsed. In the light clock of the S-frame, however, the light is already returning. Therefore, $t > 0.5$ ns must be true (Fig. 2.4). Again, we can clearly see that time runs faster in the S-frame than in the S'-frame.

With Pythagoras's theorem, t and t' are connected by $c^2 t^2 = v^2 t^2 + c^2 t'^2$. Note that ct' rather than $c't$ denotes the length of the vertical leg of the triangle! Nothing shows Einstein's ingenuity better than this choice of variables! Common sense would suggest that the speed of light is reduced in the S'-frame from c to $c' = \sqrt{c^2 - v^2}$, but the same time has elapsed in both frames. Einstein, however, postulates that the speed of light is the same in both frames which necessitates *a slower lapse of time in the primed frame*, thus giving ct' for the length in question. Rearranging the foregoing equation gives

$$t^2(c^2 - v^2) = c^2 t'^2$$

$$t^2 = \frac{t'^2}{1 - \left(\dfrac{v}{c}\right)^2}.$$

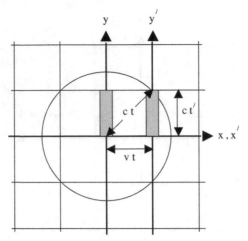

Fig. 2.4.   The light clocks at $t = 0.5\sqrt{2}$ ns.

The negative solution has no physical meaning, thus

$$t = \frac{t'}{\sqrt{1 - \left(\frac{v}{c}\right)^2}}. \qquad (2.1)$$

The square root expression $\sqrt{1 - \left(\frac{v}{c}\right)^2}$ appears many times in the special theory of relativity. Thus one defines

$$\beta = \frac{v}{c}, \qquad (2.2)$$

$$\gamma = \frac{1}{\sqrt{1 - \beta^2}}. \qquad (2.3)$$

Obviously, γ is always greater than or equal to unity. So let us summarize the results of our thought experiment:

> When a frame S' is in motion at a constant velocity v relative to another frame S, any process (e.g. the tick of a clock) being at rest in S' and requiring the time t' in S' is lengthened for an observer at rest in S to
>
> $$t = \frac{t'}{\sqrt{1 - \left(\dfrac{v}{c}\right)^2}} = \frac{t'}{\sqrt{1 - \beta^2}} = \gamma t' \geq t'.$$
>
> This phenomenon is called *time dilation*, literally *time stretching*.

These considerations remain unchanged when the origin of the S'-frame is displaced into the y- or z-direction. So the most general case, i.e. the flyby of the S'-frame past the S-frame, is already included.

Let us continue with our thought experiment: For an observer at rest in the S'-frame, the S-frame moves to the left at the velocity $v = -c/\sqrt{2}$. The same considerations show that for an observer at rest in the S'-frame a process at rest in the S-frame requiring the time t in S is lengthened to

$$t' = \gamma t \geq t. \tag{2.4}$$

This has to be so because of the principle of special relativity; failing this, one of the two reference frames would be preferred. This brings us to one of the most important conclusions:

> Time is relative. Of all inertial observers, an observer at rest relative to a process measures the shortest time for that process. This time is called the *proper time* of a process, denoted by τ.

At this point, we would like to draw the reader's attention to the coexistence of two seemingly contradictory equations that may give us a headache, namely, $t = \gamma t'$ and $t' = \gamma t$. These equations have the following meaning: The time variables on the *right-hand sides* stand for the *proper time* of a process, i.e. for the time that an observer measures who is at rest relative to that process; the left-hand sides denote the time an observer measures who is in motion relative to that process. In the equation $t = \gamma t'$, the time variable t' is the proper time of a process, the S'-frame is at rest relative to it; in the equation $t' = \gamma t$, the time variable t is the proper time of a process, the S'-frame is in motion relative to it. When the S-frame and the S'-frame are in motion relative to each other, there can only be *one* frame where a process is at rest. Therefore, one has no choice between two equations.

The question remains whether the above considerations are applicable for light clocks only, or whether they are equally true for mechanical, electric, or other clocks, or even for an aperiodic course of events, such as radioactive decay or a biological process. The answer is a clear *yes*. If there were a difference between the time scales of a light clock and another clock at rest relative to it, these differences would have to vary from one inertial reference frame to another one in order to become detectable. Then a reference frame with a minimal difference would exist. Such an inertial reference frame would be a distinguished reference frame and thus a violation of the principle of special relativity, i.e. a violation of one of the two premises of the whole theory of special relativity.

A famous experimental proof of time dilation is linked to the *lifetime of muons* which come into existence through *cosmic rays* in the *upper atmosphere*. According to the *law of radioactive decay*, $N = N_0(1/2)^{t/T_{1/2}}$, muons have a *half-life*, $T_{1/2}$, of 1.5 µs before they decay into an electron or a positron, respectively, and two neutrinos. In $t = T_{1/2} = 1.5$ µs, muons, travelling at almost the speed of light, cover a distance of 450 m. Therefore, their intensity should double with every height increase of 450 m. In reality, however, intensity increases more gradually. The reason is time dilation: For an observer on earth the rapidly moving reference frame of the muons makes time elapse more slowly, the muons' half-life is lengthened to $\gamma T_{1/2}$. The exact number is: $v = 0.994c$, or $\gamma = 9$. Thus, for an observer on earth, the muons' half-life is 14 µs within which they cover a distance of 4 km.

A rather practical problem in the age of very precise atomic clocks is the question of whether *two clocks, after having been synchronized in one place, can be moved apart without losing their synchronization*. Let us imagine that one clock, remaining at rest during the experiment, shows the time t. A second clock, being transported at the constant velocity v

over the distance s, shows the time τ. Obviously, for an observer at rest,

$$t = \frac{\tau}{\sqrt{1 - \left(\frac{v}{c}\right)^2}}. \tag{2.5}$$

Since in reality $v \ll c$ is always true, a Taylor expansion of the square root can be made, giving

$$t \approx \left[1 + \frac{1}{2}\left(\frac{v}{c}\right)^2\right]\tau,$$

or, for the time difference,

$$t - \tau \approx \frac{1}{2}\left(\frac{v}{c}\right)^2 \tau.$$

But, on the other hand, from (2.5),

$$\tau \approx \left[1 - \frac{1}{2}\left(\frac{v}{c}\right)^2\right]t$$

is also true, so the time difference between the clocks becomes

$$t - \tau \approx \frac{1}{2}\left(\frac{v}{c}\right)^2 \left[1 - \frac{1}{2}\left(\frac{v}{c}\right)^2\right]t,$$

or, to second order in v/c, simply

$$t - \tau \approx \frac{1}{2}\left(\frac{v}{c}\right)^2 t.$$

With $v = s/t$, we obtain

$$t - \tau \approx \frac{s^2}{2c^2 t}. \tag{2.6}$$

So *synchronization is lost!* However, the more time one allows for the transport over the given distance s, i.e. the more slowly the second clock is being transported, the smaller is the time difference between the two clocks.

Chapter 3*

The Hafele–Keating Experiment

In 1971, time dilation, predicted by the special (and the general) theory of relativity, was experimentally verified in a particularly convincing manner. The brilliant idea for that experiment was concocted by *Hafele and Keating*, who used scheduled flights to *carry two very precise atomic clocks around the earth* close to the equator, *one in eastbound flights and the other one in westbound flights* (Fig. 3.1).

The calculation requires two velocity variables: v_E *the velocity of the earth's rotation* at the equator (40,000 km/24 h $=$ 463 m/s), and v, *the velocity of the airplanes* carrying the atomic clocks, measured relative to the earth's surface. We now have to distinguish four reference frames:

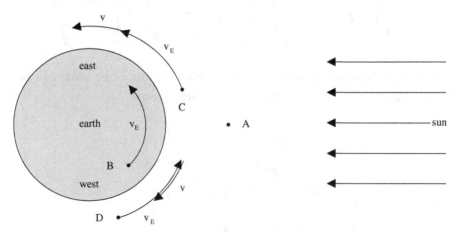

Fig. 3.1. Velocities in the Hafele–Keating experiment: v_E is the velocity of the earth's rotation, v the velocity of the airplanes.

- A is a hypothetical observer moving due west at the velocity v_E relative to the earth's surface, i.e. against the sense of the earth's rotation. For him the sun is always at its zenith, the earth is rotating underneath him. He is resting in an *inertial reference frame* because he does not participate in a circular motion around the earth's axis. The earth's revolution around the sun shall be disregarded.
- B is an observer *on the ground*; he is moving at v_E relative to A.
- C is an observer *in a plane heading east*, he is moving at $v_E + v$ relative to A.
- D is an observer *in a plane heading west*, he is moving at $v_E - v$ relative to A.

Obviously, B, C, and D are not resting in inertial frames. Since the special theory of relativity is applicable for inertial reference frames only, the calculation has to be performed in A's reference frame.

Let us idealize the situation by assuming that both planes are starting simultaneously at B and are simultaneously returning to B after having traveled once around the earth. In A's frame, the flight time is t. The proper times registered by B, C, and D, namely τ_B, τ_C, and τ_D, are functions of t as follows:

$$t = \frac{\tau_B}{\sqrt{1 - \left(\dfrac{v_E}{c}\right)^2}} = \frac{\tau_C}{\sqrt{1 - \left(\dfrac{v_E + v}{c}\right)^2}} = \frac{\tau_D}{\sqrt{1 - \left(\dfrac{v_E - v}{c}\right)^2}}.$$

Since the expressions under the roots are close to unity, the roots can be Taylor expanded. Then the three proper times become

$$\tau_B \approx t\left[1 - \frac{1}{2}\left(\frac{v_E}{c}\right)^2\right]$$

$$\tau_C \approx t\left[1 - \frac{1}{2}\left(\frac{v_E + v}{c}\right)^2\right]$$

$$\tau_D \approx t\left[1 - \frac{1}{2}\left(\frac{v_E - v}{c}\right)^2\right].$$

We calculate the difference between the proper time elapsed in the eastbound airplane and the proper time elapsed on the ground to be

$$\Delta\tau^{\text{east}} = \tau_C - \tau_B \approx -\frac{2v_E v + v^2}{2c^2}t. \tag{3.1}$$

For the westbound airplane, we calculate correspondingly

$$\Delta\tau^{\text{west}} = \tau_D - \tau_B \approx +\frac{2v_E v - v^2}{2c^2}t. \tag{3.2}$$

The airplane's velocity is assumed to be $v = 800\,\text{km/h} = 222\,\text{m/s}$; so the flight takes

$$t \approx \tau_B = \frac{40{,}000\,\text{km}}{800\,\text{km/h}} = 50\,\text{h} = 1.8 \times 10^5\,\text{s}.$$

This gives us $\Delta\tau^{\text{east}} = -255\,\text{ns}$ and $\Delta\tau^{\text{west}} = +156\,\text{ns}$. On the *eastbound flight, time does not elapse as fast as on the ground,* as expected. On the *westbound flight*, however, *time elapses somewhat faster.* This has to be so because from the viewpoint of the hypothetical inertial observer, A, the plane heading west has a lower velocity than the observer on the ground, B.

In this experiment, we also have to consider a non-negligible correction due to the *general theory of relativity* which is not the subject of this book. General relativity predicts that the lapse of time is dependent on the *gravitational potential:* The deeper a clock is positioned in the gravitational funnel of a mass the slower is its lapse of time. More details can be found at the end of Chapter 13. Since a typical cruising altitude is about 10,000 m, an extra 196 ns of time elapse in the planes in the course of 50 h when compared with the ground. Thus,

$$\Delta\tau^{\text{east}} = -255\,\text{ns} + 196\,\text{ns} = -59\,\text{ns}$$
$$\Delta\tau^{\text{west}} = +156\,\text{ns} + 196\,\text{ns} = +352\,\text{ns}.$$

From the exact flight data Hafele and Keating calculated

$$\Delta\tau^{\text{east}} = (-40 \pm 23)\,\text{ns}$$
$$\Delta\tau^{\text{west}} = (+275 \pm 21)\,\text{ns}.$$

The four atomic clocks measured

$$\Delta\tau^{\text{east}} = (-59 \pm 10)\,\text{ns}$$
$$\Delta\tau^{\text{west}} = (+273 \pm 7)\,\text{ns}.$$

Theory and experiment are in excellent agreement.

Chapter 4

Length Contraction

We use the S- and S'-reference frames from Chapter 2 for the following thought experiment: A light clock is at rest parallel to the x'-axis of the S'-frame. The S'-frame moves at a velocity of $\mathbf{v} = (v, 0, 0)$ relative to the S-frame. At the instant $t = t' = 0$, the left end of the clock is at the position $x = x' = 0$ (Fig. 4.1).

An observer in the S'-frame is igniting a light flash at $t' = 0$ and $x' = 0$ to determine the length of the light clock, ℓ', from the running time, t', of the light from the left end to the right end and back as

$$\ell' = \frac{1}{2}ct'. \tag{4.1}$$

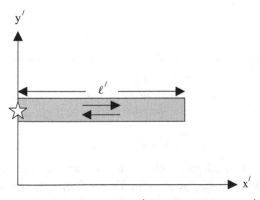

Fig. 4.1. The light clock in the S'-frame, parallel to the x'-axis.

23

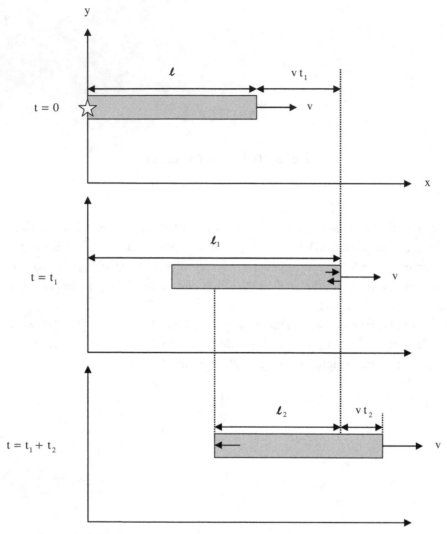

Fig. 4.2. The light clock in the S-frame moving at the velocity v ($= c/3$ in this example) into the x-direction; top: emission of a light flash at $t = 0$; center: reflection at $t = t_1$; bottom: arrival at $t = t_1 + t_2$.

In the S-frame, the light clock has the yet undetermined length ℓ. The course of the experiment, as seen in the S-frame, is shown in Fig. 4.2. Obviously, it takes the light flash more time for its way to the right end, t_1, than for its way back, t_2. As a result, the distances covered by the light are different and given by

$$\ell_1 = \ell + vt_1$$
$$\ell_2 = \ell - vt_2.$$

With $\ell_1 = ct_1$ and $\ell_2 = ct_2$, we obtain

$$(c - v)t_1 = \ell$$
$$(c + v)t_2 = \ell.$$

The running time, t, of the light flash in the S-frame is

$$t = t_1 + t_2 = \frac{\ell}{c - v} + \frac{\ell}{c + v} = \frac{\ell(c + v) + \ell(c - v)}{c^2 - v^2}$$

$$= \frac{2\ell c}{c^2 - v^2} = \frac{2\ell}{c} \frac{1}{1 - \dfrac{v^2}{c^2}} = \gamma^2 \frac{2\ell}{c},$$

thus the light clock has a length of

$$\ell = \frac{1}{\gamma^2} \frac{1}{2} ct. \tag{4.2}$$

Division of (4.1) by (4.2) gives

$$\frac{\ell'}{\ell} = \gamma^2 \frac{t'}{t}$$

$$\frac{\ell'}{\gamma \ell} = \frac{\gamma t'}{t}.$$

Since, according to (2.1) and (2.3), the right-hand side equals unity, the left-hand side must equal unity as well, whence follows

$$\ell = \frac{\ell'}{\gamma}. \tag{4.3}$$

Therefore, we conclude:

When a frame S' is in motion at a constant velocity v relative to another frame S, any length in the direction of motion being at rest in S' and amounting to ℓ' in S' is shortened for an observer at rest in S to

$$\ell = \sqrt{1 - \left(\frac{v}{c}\right)^2}\,\ell' = \sqrt{1 - \beta^2}\,\ell' = \frac{\ell'}{\gamma} \le \ell'.$$

This phenomenon is called *length contraction*, literally *length shortening*.

According to the principle of special relativity, the following is also true: When in a frame S a length ℓ is measured, it will be shortened to

$$\ell' = \frac{\ell}{\gamma} \tag{4.4}$$

in a frame S' in motion at the velocity v relative to S. By analogy to what was stated in Chapter 2 about time measurements, the equations (4.3), $\ell = \ell'/\gamma$, and (4.4), $\ell' = \ell/\gamma$, have to be interpreted in this manner: The *right-hand side* of either equation contains the *length measurement of an observer at rest* relative to the length measured.

Lengths in the direction of motion are relative. Of all inertial observers moving along a certain direction, an observer at rest relative to an object extending in that direction measures the greatest length for that object. This length is called the *proper length* of the object. Lengths in transverse directions are not subject to length contraction.

The statement about the *invariance of lengths at right angles to the direction of motion* can be proven elegantly with a thought experiment: Let us think of two hollow cylinders of equal diameter, one blue, the other one red, with their symmetry axes coinciding and moving along that axis toward each other. Let us imagine, for a moment, that there is a length contraction transverse to that axis (Fig. 4.3). In the reference frame of the blue cylinder, the *red cylinder has to shrink* and penetrate into the blue one. In the reference frame of the red cylinder, the *blue cylinder has to shrink*

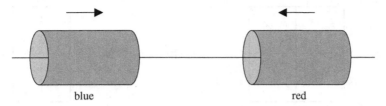

Fig. 4.3.   Two hollow cylinders on a collision course.

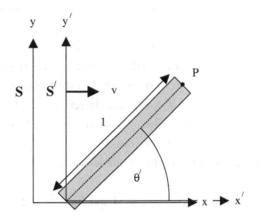

Fig. 4.4.   A rod tilted into the y-direction.

and penetrate into the red one. Both processes cannot happen at the same time, so lengths in transverse directions must remain unchanged.

Two examples from opposite ends of the velocity spectrum in the universe are apt to elucidate the phenomenon of length contraction: Let us first return to the very fast *muons* in the earth's atmosphere. This time we move with the muons in their reference frame, with the result that the muons' half-life has its proper value of $1.5 \, \mu$s, but *their flight path through the earth's atmosphere is now reduced* by a factor of $\gamma = 9$, and a distance of $4 \, $km shrinks to as little as $450 \, $m. The earth's atmosphere, racing towards the muons, is only about $10 \, $km thick, thus many muons can find their way down to the earth's surface.

Of how little importance length contraction is in present *space flight*, situated at the very low end of velocities in the universe, is shown by these considerations: At a typical velocity of $8.0 \, $km/s, or $\gamma = 1 + 3.6 \times 10^{-10}$,

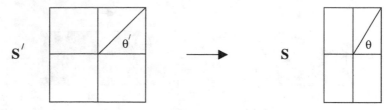

Fig. 4.5.   Distortion of a square lattice.

the earth's diameter is reduced by 4.5 mm, and even the moon is only 14 cm closer to earth.

It is also interesting to note that *angles of fast moving objects are altered* due to length contraction parallel to the direction of motion and the invariance of lengths in all transverse directions: A rod of proper length 1 in the S'-frame subtends the angle θ' with the direction of motion (Fig. 4.4). How large is θ in the S-frame? The end of the rod, P, has the coordinates $(x' = \cos\theta' \mid y' = \sin\theta')$. In the S-frame, the x'-component is subject to length contraction, but the y'-component is not, therefore $(x = \cos\theta'/\gamma \mid y = \sin\theta')$.

The tangent of θ in the S-frame is larger by a factor of γ:

$$\tan\theta = \frac{y}{x} = \gamma\tan\theta'. \qquad (4.5)$$

Thus a *square lattice* in the S'-frame turns into a *rectangular lattice* in the S-frame (Fig. 4.5). With $\beta = 0.8$, or $\gamma = 5/3$, all lengths in the direction of motion are reduced to 60% of the respective proper lengths. We will return to this problem in Chapter 16.

Chapter 5*

The Unification of Electric and Magnetic Forces

This chapter, along with others marked by an asterisk, is more advanced and can be skipped by beginners without running a risk of losing the *red thread*.

Modern theoretical physics is deeply involved with the idea of *unifying elementary forces or interactions*. This development started one hundred years ago with the special theory of relativity because it succeeded in unifying the *electric (Coulomb) force* with the *magnetic (Lorentz) force*. In other words, the electric force and the magnetic force have been identified as two aspects of *one electromagnetic force*. Depending on the reference system, it emerges as a purely electric force or as a purely magnetic force or as both. This phenomenon shall be expounded here with a sample calculation.

A *copper filament* consists of a crystal lattice of *singly charged positive copper ions* and the same number of *conduction electrons* which, to simplify matters in our model calculation, all move at the same velocity **v** to the right. This arrangement is called a *current-carrying filament*. Above the copper filament another positive charge q moves to the right at the same velocity **v**. This charge experiences a force **f** (Fig. 5.1, left). We use a lower-case letter for the force vector in order to reserve an upper-case letter for the corresponding four-vector to appear later in this book.

The S-frame just described is at rest relative to the copper ions. There is perfect *charge neutrality inside the filament*, so the Coulomb force must be ruled out as an explanation for the force. The conduction electrons, however, produce a magnetic induction **B**. Since in the S-frame the charge q above the filament moves to the right at the velocity **v**, it experiences a *Lorentz force* f_L.

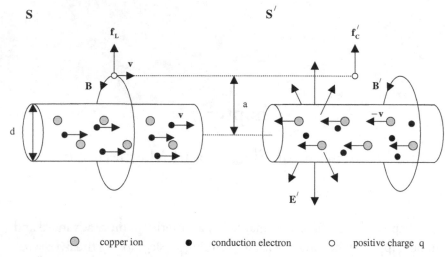

◎ copper ion	● conduction electron	○ positive charge q

Fig. 5.1.   Copper filament consisting of copper ions and conduction electrons, with one positive charge above it. Left: *S*-frame of copper ions at rest; right: *S'*-frame of conduction electrons at rest.

Let us turn to the *S'*-frame: It is at rest relative to the conduction electrons and the charge *q* above (Fig. 5.1, right). Now a magnetic induction **B'** produced by the copper ions moving to the left arises. The two fields, **B'** and **B**, are parallel. The charge above the filament does *not* experience a *Lorentz force* because its velocity in the *S'*-frame is zero. Therefore, it *must be a Coulomb force* **f'_C**; a radial electric field **E'** pointing outwards must be present. Why? In the *S'*-frame, the distances between the copper ions are subject to length contraction, so the ions are spaced more closely together. At the same time, the distances between the conduction electrons are no longer subject to length contraction, they are spaced further apart. It is for these two reasons that there is a *positive charge surplus inside the filament* thus producing a radial electric field pointing outwards.

Let us now calculate the problem in the *S*-frame: The distance of the charge *q* from the axis of the filament is *a*. According to *Ampère's law*, the magnetic induction **B** in the distance *a* from a long straight filament carrying the electric current *I* amounts to

$$B = \frac{\mu_0}{2\pi} \frac{I}{a}.$$

The *Lorentz force* f_L acting on the charge q, flying at the velocity $\mathbf{v}$ parallel to the filament, has the magnitude

$$f_L = qvB = \frac{\mu_0}{2\pi} \frac{qv}{a} I.$$

n is the number of conduction electrons per unit volume; each electron carries one elementary charge of size e. Thus ne is the charge density, i.e. the magnitude of the charge per unit volume. When all conduction electrons move at the same speed v to the right, nev is the electric current density, i.e. the amount of charge flowing per unit time through a unit of the filament's cross-section. d is the diameter of the filament. So $nev\pi(d/2)^2$ is the amount of charge flowing per unit time through the whole cross-section, i.e. the electric current I. Therefore, we find for the Lorentz force in the S-frame

$$f_L = \frac{\mu_0}{2} \frac{qv^2}{a} ne \left(\frac{d}{2}\right)^2. \tag{5.1}$$

We now want to switch from the S-frame to the S'-frame. The *conduction electrons and the charge q* located above have *zero velocity*, and the copper ions of the crystal lattice (i.e. the whole crystal lattice or the whole filament) move at the velocity $-\mathbf{v}$, i.e. to the left. The *crystal lattice is now subject to length contraction*; in the direction of the current the copper ions are spaced more closely together by a factor of $1/\gamma$. Their charge density increases from $\rho_+ = ne$ to $\rho'_+ = \gamma ne$. On the other hand, the *conduction electrons are no longer subject to length contraction* in the direction of the current, so they are further apart by a factor of γ. Their charge density is consequently reduced from $\rho_- = -ne$ to $\rho'_- = -ne/\gamma$. There is charge neutrality in the S-frame,

$$\rho = \rho_+ + \rho_- = ne - ne = 0.$$

In the S'-frame, there is a *positive charge density surplus* of

$$\rho' = \rho'_+ + \rho'_- = \gamma ne - \frac{ne}{\gamma}$$

$$= ne \left(\frac{1}{\sqrt{1-\beta^2}} - \sqrt{1-\beta^2}\right) = ne \frac{1 - (1-\beta^2)}{\sqrt{1-\beta^2}} = \gamma\beta^2 ne. \tag{5.2}$$

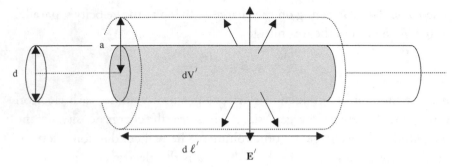

Fig. 5.2. Calculation of the integral (5.5) for the simple case of a long, straight filament.

It is obvious that primed variables belong to the S'-frame and unprimed ones to the S-frame. We will stick to this convention throughout this book.

An element of the filament of infinitesimal length $d\ell'$ and a diameter of $d' = d$ (perpendicular to the direction of motion) has a volume (Fig. 5.2) given by

$$dV' = \pi \left(\frac{d}{2}\right)^2 d\ell'.$$ (5.3)

The *positive charge surplus* inside that element amounts to

$$\rho' dV' = \gamma\beta^2 ne\pi \left(\frac{d}{2}\right)^2 d\ell'.$$ (5.4)

The electric field E' in the distance $a' = a$ from the axis of the filament can be calculated with *Gauss's law of electrostatics*,

$$\oint \mathbf{E}' \cdot \mathbf{dA}' = \frac{1}{\varepsilon_0} \int \rho' dV'.$$ (5.5)

$\mathbf{dA}'$ is a surface element; the integration on the left-hand side extends over a closed surface, e.g. the surface of a cylinder; the integration on the right-hand side yields the charge enclosed by that surface.

For the evaluation of the integral on the left-hand side, we suppose that the filament is long and straight, an infinitesimal element of which is enclosed by a coaxial cylinder of length $d\ell'$ and radius a. Everywhere, the E'-field is pointing radially outwards. The contribution of the two circular basal surface areas is zero because there $\mathbf{dA}'$ stands at right angles to $\mathbf{E}'$.

The only non-zero contribution is provided by the cylinder envelope, where E' and $\mathbf{dA}'$ are parallel, yielding

$$2\pi a d \ell' E'. \tag{5.6}$$

The right-hand side of (5.5) is simply the charge calculated in (5.4), divided by the dielectric constant of vacuum, ε_0,

$$\frac{1}{\varepsilon_0} \gamma \beta^2 n e \pi \left(\frac{d}{2}\right)^2 d\ell'. \tag{5.7}$$

The strength of the electric field E' in the distance a from the axis of the filament is obtained by equating (5.6) and (5.7),

$$E' = \frac{1}{2\varepsilon_0} \frac{1}{a} \gamma \beta^2 n e \left(\frac{d}{2}\right)^2.$$

So a charge q in the distance a is acted upon by a Coulomb force of magnitude

$$f'_C = q E' = \frac{1}{2\varepsilon_0} \frac{q}{a} \gamma \beta^2 n e \left(\frac{d}{2}\right)^2.$$

With $1/\varepsilon_0 = \mu_0 c^2$ and $\beta^2 = v^2/c^2$, we obtain

$$f'_C = \frac{\mu_0}{2} \frac{q v^2}{a} \gamma n e \left(\frac{d}{2}\right)^2. \tag{5.8}$$

A comparison of (5.1) and (5.8) yields

$$f'_C = \gamma f_L, \tag{5.9}$$

i.e. *the force which appears as a Lorentz force,* f_L, *in the S-frame is evidenced as a Coulomb force,* f'_C, *in the S'-frame*; both forces differ only by a factor of γ. The reason for this γ-factor will be explained in Chapter 33.

The derivation of (5.9) requires one annotation: In writing down (5.2) we have already made the implicit assumption that the magnitude of the

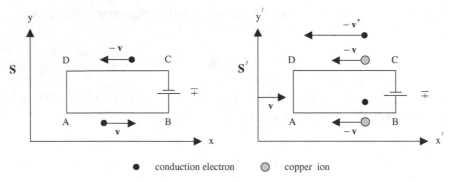

Fig. 5.3.   Closed electric circuit *ABCD*: at rest in the *S*-frame, moving at −v in the *S'*-frame.

elementary charge *e* does not depend on the reference system. This *invariance of electric charge* is, strictly speaking, *one of the postulates of special relativity*. Its experimental verification is beyond any doubt, as one can see from the following consideration: Let us think of an electrically neutral piece of metal. Heating the metal changes the speeds of both the conduction electrons and of the metal ions. The changes, however, follow different laws. If the magnitude of a charge depended on its speed, charge neutrality would be lost. No such phenomenon has ever been observed.

In particular, the sum of all positive and negative charges in the universe is zero − the *universe is electrically neutral*. This must be *true for all inertial reference frames*. Equation (5.2), however, makes us believe that the total charge, e.g. the charge of the conductor *AB* in Fig. 5.3, depends on the reference frame, i.e. is not invariant.

Since electric current can only flow in a *closed electric circuit*, there must be a return line *CD* in the opposite direction. Without altering the problem, the electric circuit may be depicted as a rectangle *ABCD*. In the *S*-frame and the *S'*-frame, electrons (small black dots) and copper ions (larger gray dots) move as shown in Fig. 5.3. In the *S*-frame there is charge neutrality. It is now left for us to show that the *S'-frame is also electrically neutral*.

First we conclude that, in switching from the *S*-frame to the *S'*-frame, positive and negative charge densities in the conductors *BC* and *AD* are raised by the same factor γ; i.e. charge neutrality is maintained. According to (5.2), a charge density surplus of $\gamma\beta^2 ne$ arises in conductor *AB*. Since

AB and *CD* are equally long, what is left to be shown is that in the *S'*-frame a charge density surplus of $-\gamma\beta^2 ne$ arises in *CD*.

In the *S*-frame, the charge density of copper ions in *CD* has the magnitude *ne* and in the *S'*-frame the somewhat higher magnitude

$$\rho'_+ = \gamma ne, \tag{5.10}$$

with γ given by

$$\gamma = \frac{1}{\sqrt{1 - \left(\dfrac{v}{c}\right)^2}}.$$

The charge density of conduction electrons in *AB* is reduced, as shown above, from $-ne$ in the *S*-frame to $-ne/\gamma$ in the *S'*-frame. In the *S'*-frame, conduction electrons in *AB* have zero velocity, in *CD* it amounts to $-v^*$. So their charge density is raised from $-ne/\gamma$ in *AB* to

$$\rho'_- = -\gamma^* \frac{ne}{\gamma}, \tag{5.11}$$

with

$$\gamma^* = \frac{1}{\sqrt{1 - \left(\dfrac{v^*}{c}\right)^2}},$$

in *CD*. At this point a result from Chapter 9 has to be anticipated. The magnitude of the electron velocity in *CD*, v^*, is not, as one might think, $v^* = v + v = 2v$, but, according to the so-called *addition theorem of velocities*,

$$v^* = \frac{v + v}{1 + \dfrac{vv}{c^2}} = \frac{2v}{1 + \left(\dfrac{v}{c}\right)^2}.$$

Thus we have

$$\gamma^* = \frac{1}{\sqrt{1 - \dfrac{4\beta^2}{(1 + \beta^2)^2}}} = \frac{1 + \beta^2}{\sqrt{(1 + \beta^2)^2 - 4\beta^2}} = \frac{1 + \beta^2}{1 - \beta^2}.$$

From (5.10) and (5.11), we conclude that in *CD* there is a charge density surplus of

$$\rho' = \rho'_+ + \rho'_- = ne\left(\gamma - \frac{\gamma^*}{\gamma}\right) = ne\left(\frac{1}{\sqrt{1-\beta^2}} - \frac{1+\beta^2}{\sqrt{1-\beta^2}}\right) = -\gamma\beta^2 ne,$$

Q.E.D.

Chapter 6

Lorentz Transformation

The quantitative treatment of problems in special relativity necessitates two inertial reference frames: the S-frame with the orthogonal coordinates x, y, and z, and the time coordinate t, and the S'-frame with the orthogonal coordinates x', y', and z', along with the time coordinate t'. The three spatial coordinate axes of these frames are mutually parallel, with S' moving at a constant velocity of $\mathbf{v} = (v, 0, 0)$ relative to S. The origins of both frames coincide at $t = t' = 0$ (Fig. 6.1).

An *event* is any *point in the four-dimensional space-time* continuum, given by four numbers for the variables x, y, z, and t. An event may be very spectacular like a supernova explosion or quite unspectacular like reading

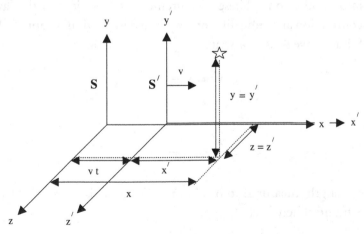

Fig. 6.1. An event in the S-frame and in the S'-frame.

these lines here and now. Once the coordinate origin ($x = y = z = 0$) and the beginning of the time scale ($t = 0$) have been defined for the S-frame, and once the velocity v, governing the motion of the S'- relative to the S-frame, has been laid down, one may want to calculate the space-time coordinates x', y', z', and t' of that event in the S'-frame.

At first glance, the problem seems to be quite trivial because the conversion of x into x' appears to be given by $x' = x - vt$, and the y-, z- and t-coordinates appear to remain unchanged. In order to give all four coordinates the dimension of length, the time coordinate, t, is multiplied by the speed of light, c. This coordinate transformation is called the *Galilei transformation* or *Galilei transform*,

$$ct' = ct$$
$$x' = x - vt$$
$$y' = y$$
$$z' = z. \tag{6.1}$$

On closer inspection, however, we find a serious mistake. Length measurements in a direction parallel to the relative velocity, v, as well as time measurements cannot simply be transferred from one inertial reference frame into another one. These measurements are subject to the laws of length contraction and time dilation, explained in previous chapters. Thus, in the S-frame, we must *not* write

$$x' = x - vt,$$

but

$$\frac{x'}{\gamma} = x - vt$$

because a length, measured to be x' in the S'-frame, is contracted to x'/γ in the S-frame. Therefore,

$$x' = \gamma(x - vt), \tag{6.2}$$

or

$$x' = \gamma(x - \beta ct).$$

In the S'-frame, correspondingly, the distance between the two coordinate origins is given by vt', *not* by vt. Because of the principle of special relativity, i.e. the equivalence of the two inertial frames, the relative velocity, v, must have the same magnitude in both frames. Thus, in the S'-frame, we must *not* write

$$x' = x - vt',$$

but

$$x' = \frac{x}{\gamma} - vt' \tag{6.3}$$

because a length, measured to be x in the S-frame, is contracted to x/γ in the S'-frame. Equating (6.2) and (6.3) yields

$$\gamma x - \gamma vt = \frac{x}{\gamma} - vt'$$
$$vt' = \gamma \left[vt + x \left(\frac{1}{\gamma^2} - 1 \right) \right]$$
$$vt' = \gamma(vt - \beta^2 x)$$
$$ct' = \gamma(ct - \beta x). \tag{6.4}$$

Since the relative motion has no y- and z-components, these components remain unchanged. We summarize,

$$ct' = \gamma(ct - \beta x)$$
$$x' = \gamma(x - \beta ct)$$
$$y' = y$$
$$z' = z. \tag{6.5}$$

This set of equations is called *Lorentz transformation* or *Lorentz transform*, named after its discoverer. It represents the relativistic generalization of the Galilei transformation. The epoch-making difference, when compared with the Galilei transformation (6.1), is the *inequality of the times involved*: t' in the S'-frame is *not equal* to t in the S-frame. Time t' depends on *both* time t *and* position x, as much as position x' depends on *both* x and t. A novel *symmetry between time and position, or space*, emerges.

For very small velocities when compared with the speed of light, β approaches 0 and therefore γ approaches 1. Then, as one can easily see, the Lorentz transformation is reduced to a Galilei transformation. The arithmetic accuracy required for a certain problem determines the minimum velocity below which the Lorentz transformation may be replaced by the Galilei transformation. As has already been shown in previous chapters, velocities in contemporary aeronautics and space flight, even an interplanetary one, are still so minute that non-relativistic calculations yield excellent results.

What is left to be shown is the *inverse Lorentz transformation*, i.e. the calculation of coordinates in the S-frame when coordinates of an event in the S'-frame are known. For this purpose, one interchanges the S-frame with the S'-frame; i.e. the new S'-frame moves at the velocity $-v$ parallel to the x-axis. In mathematical terms, all primed variables have to be replaced by unprimed ones and vice versa, and v has to be replaced by $-v$, thus $\beta \rightarrow -\beta$ and $\gamma \rightarrow \gamma$. The result is

$$
\begin{aligned}
ct &= \gamma(ct' + \beta x') \\
x &= \gamma(x' + \beta ct') \\
y &= y' \\
z &= z'.
\end{aligned}
\tag{6.6}
$$

The inverse Lorentz transformation can also be derived from (6.5). One multiplies the second line of (6.5) by β and adds the first line,

$$
\begin{aligned}
ct' + \beta x' &= \gamma ct - \gamma \beta^2 ct \\
ct' + \beta x' &= \gamma ct \frac{1}{\gamma^2} \\
ct &= \gamma(ct' + \beta x').
\end{aligned}
$$

Correspondingly, one multiplies the first line of (6.5) by β and adds the second line,

$$\beta ct' + x' = -\gamma\beta^2 x + \gamma x$$

$$\beta ct' + x' = \gamma x \frac{1}{\gamma^2}$$

$$x = \gamma(x' + \beta ct').$$

Before we continue, one short comment is required on the equations of time dilation (2.1 and 2.4) and length contraction (4.3 and 4.4), as opposed to the Lorentz transformation formulas (6.5 and 6.6). These equations are frequently misunderstood! (6.5) and (6.6) are formulas for the transformation of coordinates, i.e. *points in the space-time continuum*, while (2.1; 2.4) and (4.3; 4.4) are relations between *certain time intervals* and *lengths*.

As a *test of consistency* for the Lorentz transformation we would like to apply it in order to re-derive time dilation and length contraction.

Time dilation: Two events are happening in the S'-frame at the location (x', y', z') and at the times t'_1 and t'_2. What temporal spacing do they have in the S-frame? One takes the first line of (6.6) twice and subtracts them,

$$c(t_2 - t_1) = \gamma[c(t'_2 - t'_1) + \beta(x' - x')]$$

$$t_2 - t_1 = \gamma(t'_2 - t'_1).$$

Of course, the (inverse) Lorentz transformation must also contain the information that time dilation is inverted when the S- and S'-frames are interchanged: Two events are happening in the S-frame at the location (x, y, z) and at the times t_1 and t_2. Their temporal spacing in the S'-frame is calculated by taking the first line of (6.5) twice, followed by subtraction,

$$c(t'_2 - t'_1) = \gamma[c(t_2 - t_1) - \beta(x - x)]$$

$$t'_2 - t'_1 = \gamma(t_2 - t_1).$$

Length contraction: In the S'-frame a rod at rest extends from x'_1 to x'_2. Its length in the S-frame is calculated by taking the second line of (6.5) twice, followed by subtraction,

$$x'_2 - x'_1 = \gamma[(x_2 - x_1) - \beta c(t_2 - t_1)].$$

A length measurement in the S-frame, where the rod is in motion, is a reasonable procedure only if the locations of both ends of the rod, x_1 and x_2, are determined at the same time, $t_1 = t_2$. This gives us

$$x_2 - x_1 = \frac{1}{\gamma}(x_2' - x_1'). \tag{6.7}$$

Let us again interchange the S- and S'-frames, starting with a rod at rest extending from x_1 to x_2, whose length is to be determined in the S'-frame. One takes the second line of (6.6) twice and subtracts them,

$$x_2 - x_1 = \gamma[(x_2' - x_1') + \beta c(t_2' - t_1')].$$

Since a length measurement in the S'-frame is reasonable only if x_1' and x_2' are determined at the same time, $t_1' = t_2'$, we obtain

$$x_2' - x_1' = \frac{1}{\gamma}(x_2 - x_1). \tag{6.8}$$

At this point one might ask why the Lorentz transformation (6.5) is applied for the above derivation of the length contraction (6.7) instead of the inverse Lorentz transformation (6.6), since the primed coordinates are given and the unprimed ones are solved for. A similar argument could be advanced for the derivation of (6.8).

The solution of this "enigma" is quite simple: The Lorentz transformation (6.5) and the inverse Lorentz transformation (6.6) contain *exactly the same information*, as (6.6) can be derived exclusively from (6.5), as has been shown above. Whether (6.5) or (6.6) is used for the solution of a particular problem depends only on which of these equations is more practical at the moment. The derivation of (6.7) by means of (6.5) is certainly a more practical choice as the time difference disappears.

So let us demonstrate the more tedious derivation of (6.7) from (6.6): We take the second line of (6.6) twice and subtract,

$$x_2 - x_1 = \gamma[(x_2' - x_1') + \beta(ct_2' - ct_1')].$$

We do *not* necessarily have $t_1' = t_2'$, as the rod is at rest in the S'-frame and the x_1'- and x_2'-coordinates remain unchanged at all times t'. To get rid of the time coordinates the first line of (6.6) has to be solved for ct',

$$ct' = \frac{ct}{\gamma} - \beta x',$$

and inserted above,

$$x_2 - x_1 = \gamma \left[(x_2' - x_1') + \beta \left(\frac{ct_2}{\gamma} - \beta x_2' - \frac{ct_1}{\gamma} + \beta x_1' \right) \right].$$

Since $t_1 = t_2$ holds, we obtain

$$x_2 - x_1 = \gamma [(1 \quad \beta^2)(x_2' - x_1')]$$

$$x_2 - x_1 = \frac{1}{\gamma}(x_2' - x_1'),$$

Q.E.D.

In many textbooks the Lorentz transformation is derived from more general principles. These ideas are rather demanding and may be skipped during the first perusal of this book. The alternative derivation of the Lorentz transformation is largely of a mathematical nature and starts with the realization that the coordinates ct', x', y', and z' must be written as *linear functions* of ct, x, y, and z, i.e. by means of the following transformation:

$$ct' = a_{00}ct + a_{01}x + a_{02}y + a_{03}z$$

$$x' = a_{10}ct + a_{11}x + a_{12}y + a_{13}z$$

$$y' = a_{20}ct + a_{21}x + a_{22}y + a_{23}z$$

$$z' = a_{30}ct + a_{31}x + a_{32}y + a_{33}z. \qquad (6.9)$$

The inverse transformation must also be linear, namely

$$ct = a_{00}'ct' + a_{01}'x' + a_{02}'y' + a_{03}'z', \qquad (6.10)$$

etc. All coefficients may be functions of the relative velocity v.

That transformation and inverse transformation look alike, i.e. that they are functions of the same type, is demanded by the principle of *shape invariance* (Chapter 1). Only linear functions can fulfill this requirement. To give an example, the inverse of the square function is the square root function; the exponential function's inverse is the logarithmic function etc. Non-linear functions would have the additional disadvantage that certain points are particular, such as minima, maxima, singularities etc. All this would violate the idea that *space is homogeneous*, i.e. that space has equal qualities everywhere.

The coefficients a_{02} and a_{03} must be zero because otherwise two events symmetric to the x-axis, e.g. at $y = y_0$ and $y = -y_0$, would take place at different times in the S'-frame. This would violate the idea that *space is isotropic*, i.e. that the qualities of space have no directional dependence. The isotropy also requires $a_{20} = a_{30} = 0$, otherwise points such as the origin of the S-frame ($x = y = z = 0$) would move away from the x'-axis ($y', z' \neq 0$), and a certain direction would be distinguished. Thus *homogeneity and isotropy of space* enter explicitly into the special theory of relativity.

The coefficients $a_{12}, a_{13}, a_{21}, a_{23}, a_{31}$, and a_{32} must also be zero. Let us take, for example, the x'-z'-plane intersecting the y'-axis at $y' = 0$. It is defined by $y' = 0$ and $x', z' \in \Re$. If, for example, a_{21} differed from zero, $y' = 0$ would only be satisfied with a_{22} and/or a_{23} being unequal to zero as well. The plane thus spanned would not be parallel to the x-z-plane, and the axes of the two frames would no longer be parallel.

Now let us turn to the coefficients a_{22} and a_{33}: They must be equal in order not to distinguish a certain direction (y, z). The transformation and the inverse transformation for the y-direction read $y' = a_{22}(v)y$ and $y = a'_{22}(-v)y'$ respectively, because, as outlined earlier, primed and unprimed coordinates have to be interchanged and v replaced by $-v$. Inserting one equation into the other one yields $a_{22}(v)a'_{22}(-v) = 1$. The shape invariance of the Lorentz transformation requires that $a_{22}(v)$ and $a'_{22}(-v)$ have to be expressed by the same function. But only the function $a_{22}(v) = a'_{22}(-v) = \text{const}$ can satisfy this condition. Therefore, $a_{22}^2 = 1$ and $a_{22} = \pm 1$. The negative sign is excluded because for $v \to 0$ the already known result must be obtained, i.e. $y = y'$. (6.9) is thus simplified to

$$ct' = a_{00}ct + a_{01}x$$
$$x' = a_{10}ct + a_{11}x$$
$$y' = y$$
$$z' = z. \tag{6.11}$$

A spherical wave expanding into space from the common origin of two frames at $t = t' = 0$ can be expressed in the S-frame and the S'-frame, respectively, by

$$x^2 + y^2 + z^2 = (ct)^2$$

and

$$x'^2 + y'^2 + z'^2 = (ct')^2,$$

where ct and ct' are the respective radii. Thus,

$$x^2 + y^2 + z^2 - (ct)^2 = 0 = x'^2 + y'^2 + z'^2 - (ct')^2.$$

Since $y = y'$ and $z = z'$, this equation is reduced to

$$x^2 - (ct)^2 = x'^2 - (ct')^2. \tag{6.12}$$

The transformation (6.11) must satisfy (6.12), so we insert (6.11) and obtain

$$x^2 - (ct)^2 = (a_{10}ct + a_{11}x)^2 - (a_{00}ct + a_{01}x)^2$$
$$= (a_{10}^2 - a_{00}^2)(ct)^2 + (a_{11}^2 - a_{01}^2)x^2 + (2a_{10}a_{11} - 2a_{00}a_{01})ctx.$$

A comparison of coefficients gives

$$a_{00}^2 - a_{10}^2 = 1$$
$$a_{11}^2 - a_{01}^2 = 1$$
$$a_{10}a_{11} - a_{00}a_{01} = 0. \tag{6.13}$$

These are three equations with four unknowns. The fourth equation is obtained by stating that the origin of the S-frame ($x = y = z = 0$) is described in the S'-frame by $x' = -vt'$ if the axes of both frames coincide at $t = t' = 0$. Insertion into (6.11) yields

$$ct' = a_{00}ct$$
$$-vt' = a_{10}ct,$$

or

$$a_{10} = -\frac{v}{c}a_{00}. \qquad (6.14)$$

The system of equations in (6.13) to (6.14) can be solved directly. The result is $a_{00} = a_{11} = \gamma$ and $a_{01} = a_{10} = -\beta\gamma$.   Q.E.D.

Chapter 7

Minkowski Diagrams

"Henceforth space by itself and time by itself are doomed to fade away into mere shadows, and only a kind of union of the two will preserve an independent reality." (Minkowski 1908; translated by Rindler 2001, p. 89)

The *union of space and time*, though often imbued with mystery, can be portrayed very concisely and impressively by means of diagrams which illustrate the essence of the Lorentz transformation. First, however, *oblique coordinates* have to be introduced.

Normally, the position of a point P in a plane is expressed by *Cartesian (or parallel or rectangular) coordinates*, x_0 and y_0 (dashed in Fig. 7.1). But one might just as well use two oblique-angled axes, x' and y', instead. In order to determine the coordinates x_0' and y_0' of a point P in such a coordinate system, parallels to the oblique axes have to be drawn through P (dotted in Fig. 7.1), just to read off x_0' and y_0' at their intersections with the x'- and y'-axes. As a further generalization, the units of length on the new axes need not be the same as those on the old ones. In Fig. 7.1 for example, we read the values $(x_0 \approx 3.37 | y_0 \approx 3.31)$ and $(x_0' \approx 1.26 | y_0' \approx 1.21)$.

The Lorentz transformation is an *affine transformation*, i.e. a one-to-one (or bijective) mapping of a plane on itself that preserves the rectilinearity and parallelism of straight lines, as it has the form

$$ct' = a_{00}ct + a_{01}x + a_0$$
$$x' = a_{10}ct + a_{11}x + a_1.$$

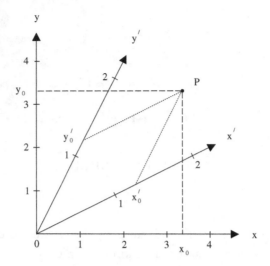

Fig. 7.1.   Oblique coordinates.

Since $a_0 = a_1 = 0$, the position of the origin remains unchanged. Now let us calculate the directions and the units of length of the new (primed) axes in the coordinate system of the old (unprimed) axes. To do so we insert the event with the coordinates $(ct' = 0 \mid x' = 1)$ into the inverse Lorentz transformation (6.6) and obtain $(ct = \beta\gamma \mid x = \gamma)$. We also insert the event $(ct' = 1 \mid x' = 0)$ to obtain $(ct = \gamma \mid x = \beta\gamma)$. Thus, the ct'- and x'-coordinate axes are located in the ct-x-coordinate system as shown in Fig. 7.2.

The usual choice of axes (horizontal: time, vertical: position) has traditionally been reversed in Minkowski diagrams. In order to not confuse readers who also browse through other books of relativity, we shall follow this convention. Note that a Lorentz transformation does not cause the coordinate axes to *rotate unidirectionally* (in the same sense), as with the rotation of a coordinate frame around its origin, but rather *counterdirectionally* (in the opposite sense). The angle between the primed and the unprimed axes is given by

$$\tan \delta = \beta = \frac{v}{c}. \tag{7.1}$$

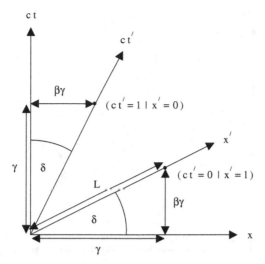

Fig. 7.2. Coordinate axes in the S- and in the S'-frame.

The unit of length, L, on the new axes can be calculated using the Pythagoras theorem to be

$$L = \sqrt{\gamma^2 + \beta^2 \gamma^2}$$

$$L = \sqrt{\frac{1 + \beta^2}{1 - \beta^2}} \qquad (7.2)$$

Let us calculate an example: The relative velocity between the S- and the S'-frames is $v = c/2$. This gives $\tan \delta = \beta = 1/2$ and $L = \sqrt{5/3} \approx 1.29$. Therefore, upon turning from the old to the new set of coordinate axes, the spacing of the tick marks is stretched by a factor of 1.29. Let us set the units of length on the x- and x'-axes to be one light-year (ly). For the sake of uniformity in dimensions, the t- and t'-axes are shown as ct- and ct'-axes, so that their units of length are also one light-year (Fig. 7.3). Finally, to avoid confusion among beginners, we want to remark that the obliquity of the x'-axis *in a Minkowski diagram* does not mean that there is an angle between the x- and x'-axes *in space* (see also Fig. 6.1).

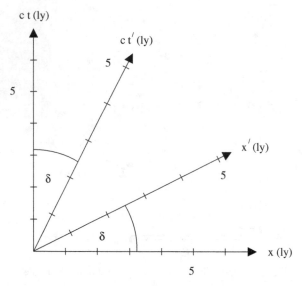

Fig. 7.3.   Units of length on the coordinate axes.

Some examples will demonstrate the power of Minkowski diagrams to unravel transformations between space-time coordinate frames.

Example 1: Events

Example (a), (Fig. 7.4): Event E is located 5 years in the future and 3 light-years on the right of the origin, as given in the S-frame. The S'-frame moves at a velocity of $\beta = 0.5$ (i.e. to the right). When and where does event E happen in the S'-frame?

Calculation:

$$\beta = 0.5 \Rightarrow \gamma = \frac{2}{\sqrt{3}}$$

$$ct' = \gamma\,(ct - \beta x) = \frac{2}{\sqrt{3}}\,(5\,\mathrm{ly} - 0.5 \times 3\,\mathrm{ly}) \approx 4.04\,\mathrm{ly}$$

$$x' = \gamma\,(x - \beta ct) = \frac{2}{\sqrt{3}}\,(3\,\mathrm{ly} - 0.5 \times 5\,\mathrm{ly}) \approx 0.58\,\mathrm{ly}.$$

Thus we find for S': The event is about 4.04 years in the future and about 0.58 light-years on the right of the origin.

Construction:

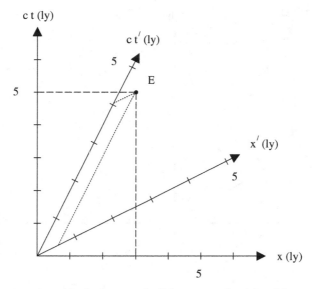

Fig. 7.4. An event 5 years in the future and 3 light-years on the right of the origin (*S*-frame).

Example (b), (Fig. 7.5): Event *E* is located 2 years in the past and 2 light-years on the left of the origin, as given in the *S*-frame. The *S'*-frame moves at a velocity of $\beta = -0.6$ (i.e. to the left). When and where does event *E* happen in the *S'*-frame?

Calculation:

$$\beta = -0.6 \Rightarrow \gamma = \frac{5}{4}$$

$$ct' = \gamma\,(ct - \beta x) = \frac{5}{4}\left(-2\,\mathrm{ly} - (-0.6)(-2\,\mathrm{ly})\right) = -4\,\mathrm{ly}$$

$$x' = \gamma\,(x - \beta ct) = \frac{5}{4}\left(-2\,\mathrm{ly} - (-0.6)(-2\,\mathrm{ly})\right) = -4\,\mathrm{ly}.$$

In *S'*, the event is located 4 years in the past and 4 light-years on the left of the origin.

Construction:

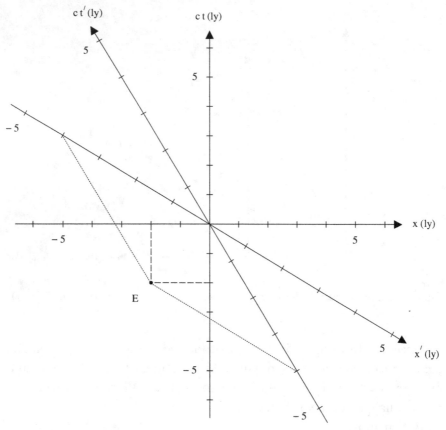

Fig. 7.5.   An event 2 years in the past and 2 light-years on the left of the origin (*S*-frame).

Example 2: Simultaneity ($\beta = 0.5$)

All events located on a *straight line parallel to the x-axis* (dashed in Fig. 7.6) are *simultaneous in S*. Consequently, all events located on a *straight line parallel to the x'-axis* (dotted) are *simultaneous in S'*. Obviously, two events, E_1 and E_2, that are simultaneous in one inertial reference frame cannot be simultaneous in any other inertial reference frame. This paradoxical situation will be the topic of the next chapter.

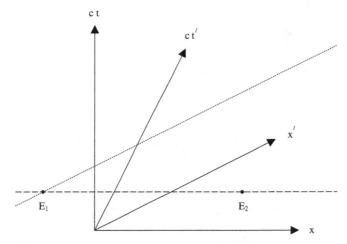

Fig. 7.6.   Simultaneity in a Minkowski diagram.

Example 3: World Lines

Let us choose any inertial reference frame, e.g. the S-frame, and locate a particle in space ($x = x_0$) and time ($ct = ct_0$). This point of space-time shall be called the event E_0. One moment later, the particle is at $x = x_1$ and $ct = ct_1$, which shall be called the event E_1. We keep track of the particle through space-time, thus obtaining a whole sequence of events. This *sequence* is called the *world line* of the particle (Fig. 7.7).

It is immediately obvious that the world line of this particle is *nothing but a path-time diagram*, with the path plotted to the right and time plotted towards the top. The world line of a particle at rest in the S-frame is parallel to the ct-axis. The faster the particle is in motion in the S-frame the flatter is its world line.

For a particle moving at the speed of light, the world line is a straight line with a slope of $+1$ or -1 (Fig. 7.8), as it moves one unit in the x-direction (1 light-year) in one unit of the ct-direction ($c \times 1$ year). As will be shown at the end of this chapter and in later chapters, no material object can move faster than the speed of light. World lines intersecting the x-axis at angles less than $45°$ are thus excluded as a matter of principle.

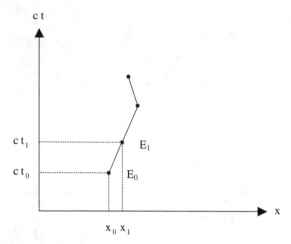

Fig. 7.7.   World line of a particle.

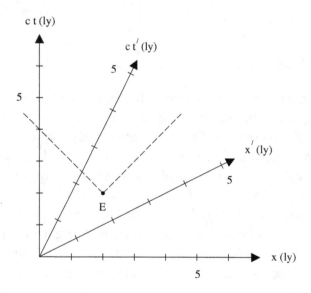

Fig. 7.8.   Propagation of a light flash in a Minkowski diagram.

Let us now investigate a light flash propagating from the event $E\,(ct = 2\,\text{ly}\,|\,x = 2\,\text{ly})$ as a spherical shell expanding in all directions. In a one-dimensional representation of space, as in Fig 7.8, the spherical shell is, at any one moment, reduced to two points (intersection of a line with a spherical shell). As time passes, the two points turn into two diverging world lines (dashed lines in Fig. 7.8). One can read directly from the diagram that the light reaches the S-observer at $t = 4$ years, and the S'-observer ($\beta = 0.5$) at $t' \approx 2.3$ years.

Example 4: Time Dilation ($\beta = 0.5$)

The point P' ($ct' = 3$ ly | $x' = 0$ ly) of the S'-frame (with oblique coordinate axes) has the time coordinate $ct \approx 3.5$ ly in the Cartesian S-frame, as indicated by the dotted auxiliary line in Fig. 7.9. This means: A process, extending from the origin to point P', that takes 3 years in the S'-frame will take about 3.5 years in the S-frame. The S-observer concludes that *the clocks of the S'-observer are slower* than his.

The point Q ($ct = 3$ ly|$x = 0$ ly) of the Cartesian S-frame has the time coordinate $ct' \approx 3.5$ ly in the oblique S'-frame, as indicated by the dashed auxiliary line in Fig. 7.9. This means: A process, extending from the origin to point Q, that takes 3 years in the S-frame will take about 3.5 years in the S'-frame. The S'-observer concludes that *the clocks of the S-observer are slower* than his.

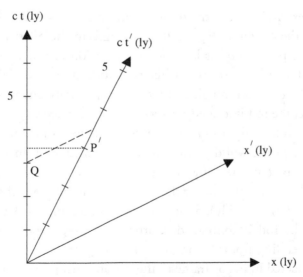

Fig. 7.9. Time dilation in a Minkowski diagram.

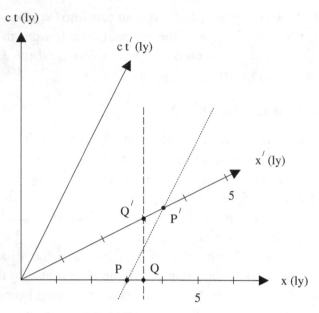

Fig. 7.10. Length contraction in a Minkowski diagram.

Example 5: Length Contraction ($\beta = 0.5$)

A rod, extending from the origin to point P' ($ct' = 0\,\text{ly} \mid x' = 3.5\,\text{ly}$), is at rest in the S'-frame (Fig. 7.10). Its length in the S'-frame amounts to 3.5 ly. The position of its left end as a function of time, i.e. its world line, coincides with the ct'-axis, while the world line of its right end is a parallel to the ct'-axis through P' (dotted line). At any one moment t' in the S'-frame, the rod is located *on a parallel to the x'-axis*.

In the S-frame, the rod *also* extends from the world line of the left end (ct'-axis) to the world line of the right end (dotted line). However, at any moment t in the S-frame, the rod is located *on a parallel to the x-axis*. At the moment $t = 0$, the rod is extending from the origin to point P ($ct = 0\,\text{ly} \mid x \approx 3\,\text{ly}$). So its length is approximately 3 ly instead of 3.5 ly, and the rod is contracted. Correspondingly, a rod at rest in the S-frame, extending from the origin to point Q ($ct = 0\,\text{ly} \mid x = 3.5\,\text{ly}$), will be contracted in the S'-frame to the distance from the origin to point Q' ($ct' = 0\,\text{ly} \mid x' \approx 3\,\text{ly}$).

At this point of our discussion it becomes clear that the phenomenon of length contraction has to do with simultaneity (see Example 2): Length measurements of moving objects are reasonable only if the positions of both ends can be measured at the same time. Since S- and S'-observers do not agree on simultaneity, they cannot agree on the results of their length measurements.

Example 6: Past, Future, Causality

Figure 7.11 contains the coordinate axes of three reference frames (S, S' ($\beta = -0.6$), and S'' ($\beta = 0.6$)), and also the two world lines of light rays passing through the origin (dashed bisectors). The projections of the events P, Q, and R on the three time axes of the S-, S'- and S''-frames are dotted.

One can immediately see: event P *is located in the future* (S'-frame: $ct' > 0$), *in the present* (S-frame: $ct = 0$), *and in the past* (S''-frame: $ct'' < 0$) respectively. Q *is located in the future of all frames, and R is located in the past*

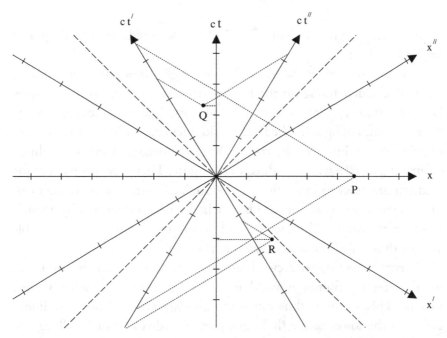

Fig. 7.11.   Future, present, and past in a Minkowski diagram.

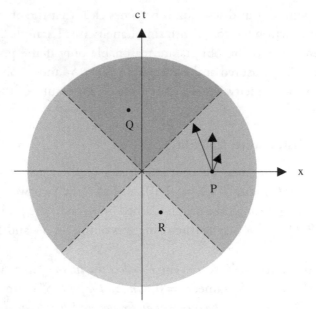

Fig. 7.12.   Future, present, and past in a Minkowski diagram again.

of all frames. For greater clarity, Fig. 7.11 has been redrawn in a simplified way in Fig. 7.12.

The two sectors on the left and on the right (*medium-gray*) comprise areas which, depending on the inertial reference frame, can be compre-hended as *future, present, or past.* Since all world lines are at least as steep as the world lines of light (dashed), all world lines originating from any event therein have to intersect the *ct*-axis above the origin. A few world lines, originating in *P*, have been added to the figure. Therefore, no event in the medium-gray sectors can influence the origin. Correspondingly, no event in the medium-gray sectors can be reached, or be influenced, by a world line from the origin. These sectors and the origin are mutually inaccessible and are thus called the *absolute alibi* of the origin.

Events in the upper sector (dark gray) are regarded as future by all inertial reference frames. A world line from the origin can reach any event therein. This sector is thus called the *absolute future.* Correspondingly, events in the lower sector (light gray) are regarded as past by all frames. Any event therein can be connected to the origin by a world line. This sector is thus called the *absolute past.*

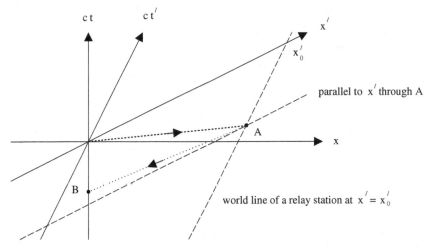

Fig. 7.13. Signal travelling at 10 times the speed of light from the origin to *B* via *A* (dotted).

Of course, it is in principle impossible for a world line to run from the upper to the lower sector, i.e. for event *Q* to influence event *R*. By no means can *Q* be the cause of *R*. However, *R* can be the cause of *Q*. The *sequence of cause and effect, causality,* cannot be inverted by the special theory of relativity. This would be a very strong argument against the theory.

Example 7: Speed Faster than Light ($\beta = 0.5$)

Let us suppose information can be transmitted at a speed faster than light (Fig. 7.13). Then one could think of the following experiment: A transmission tower is at rest in the origin of the *S*-frame, a relay station is at rest in the *S'*-frame at the position $x' = x'_0$. At the time $t = t' = 0$, a signal is being transmitted from the common origin of the two frames to the relay station at a speed of $10\,c$, as measured in *S* (square dots). It is being received by the relay station in the event *A*, and being re-emitted after a short period of time to the source at a speed of $-10\,c$, as measured in *S'* (round dots). Finally, the signal is being received in the event *B* at a time $t < 0$, i.e. at the source, but before the signal was emitted there. If the signal carries a destructive energy, it can destroy the transmitter. Cause and effect would now happen in an inverted sequence.

Chapter 8

Simultaneity

Chapter 7 introduced the idea that even *simultaneity is a relative concept*: When two events, E_1 and E_2, take place simultaneously in one inertial reference frame, they do not do so in any other inertial reference frame. This chapter is intended to further demonstrate these circumstances geometrically and arithmetically.

Let us start with Einstein's definition of simultaneity: *Two events, E_1 and E_2, taking place at two different locations, are considered simultaneous if two spherical light waves, emitted with the events, meet each other at the center of the tie line connecting the locations of the events.* This definition seems evident. Let us see how it applies to the events E_1 and E_2 of Fig. 8.1.

Two light flashes, emitted by events E_1 and E_2, meet each other in the event F. Clearly, E_1 and E_2 are *not* simultaneous in the S-frame because the projections of the line segments E_1F and FE_2 onto the x-axis are *not* equally long, as indicated by the vertical dotted lines. The lack of simultaneity of the two events in the S-frame is immediately obvious from their different time coordinates. However, when an S'-frame is constructed in such a way that the x'-axis is parallel to the line segment E_1E_2, the events E_1 and E_2 must be simultaneous in that frame. Projections of the line segments E_1F and FE_2 onto the x'-axis, indicated by oblique dotted lines, also suggest this. A geometric proof is straightforward:

The line E_1G is parallel to the x'-axis, and the line GF is parallel to the ct'-axis; E_1F, being a world line of light, has a slope of one. Thus the angles GE_1F and E_1FG are equal, and the triangle E_1GF is isosceles, i.e. the line segments E_1G and GF are equally long. Similar reasoning shows

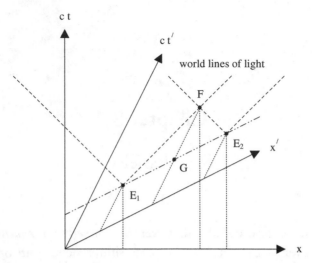

Fig. 8.1. (Lack of) simultaneity of two events, E_1 and E_2, in a Minkowski diagram.

that the triangle FGE_2 is isosceles, i.e. the line segments FG and GE_2 are equally long. Thus, the line segments E_1G and GE_2 are also equally long; the x'-coordinate of F is halfway between the x'-coordinates of E_1 and E_2. Q.E.D.

Arithmetic proof: In the S'-frame, two events, E_1 ($ct' = ct_1' \mid x' = x_1'$) and E_2 ($ct' = ct_2' \mid x' = x_2'$), are to be simultaneous, i.e. $t_1' = t_2'$. So we obtain from the inverse Lorentz transformation, (6.6),

$$ct_1 = \gamma(ct_1' + \beta x_1')$$
$$ct_2 = \gamma(ct_2' + \beta x_2')$$
$$c(t_2 - t_1) = \gamma\beta(x_2' - x_1')$$

$$t_2 - t_1 = \frac{\beta}{c}\gamma(x_2' - x_1') = \frac{1}{\sqrt{1 - \left(\dfrac{v}{c}\right)^2}}\frac{v}{c^2}(x_2' - x_1'). \qquad (8.1)$$

Two events which are simultaneous in the S'-frame are simultaneous in the S-frame as well ($t_1 = t_2$) if they have *the same position in the S'-frame* ($x_1' = x_2'$) and/or *the relative velocity v between the frames is zero*.

The corollary is: The larger the spatial separation is between two simultaneous events in the S'-frame, and/or the higher the relative velocity is between the frames, the greater is the temporal separation of these events in the S-frame.

The simultaneity of two events is relative, i.e. it depends on the inertial reference frame.

Let us continue with one more derivation of equation (8.1) by means of a thought experiment: A light clock of proper length ℓ' is at rest in the S'-frame. Its x'-axis moves at the velocity v parallel to the x-axis of the S-frame. The mirrors M are attached to the short ends of the clock, and light propagates parallel to the direction of motion. At $t = t' = 0$, when the center of the light clock ($x' = 0$) coincides with the origin of the S-frame ($x = 0$), a light flash is emitted at $x = x' = 0$ (Fig. 8.2).

For an observer in the S'-frame, light propagates in both directions at the speed of light, c, reaching both mirrors at the same time, $t' = \ell'/2c$. For an S-observer, the length of the light clock is contracted to $\ell = \sqrt{1 - (v/c)^2}\,\ell'$, but light still propagates at a speed of c in both directions. However, the *left-hand mirror is moving toward the light* at a speed of v. In the time interval t_1, required for light to reach the left mirror, light does not cover $\ell/2$, but $\ell/2$ reduced by vt_1, so the total distance covered is

$$ct_1 = \frac{\ell}{2} - vt_1.$$

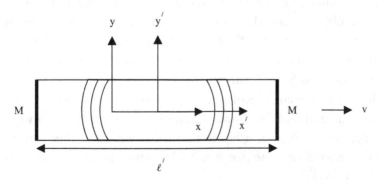

Fig. 8.2. Thought experiment with a light clock.

Correspondingly, *the right-hand mirror is running away*. In the time interval t_2, required for light to reach it, light has to cover the extra distance vt_2, i.e. a total distance of

$$ct_2 = \frac{\ell}{2} + vt_2.$$

Both equations combined yield

$$t_2 - t_1 = \frac{\ell}{2(c-v)} - \frac{\ell}{2(c+v)} = \frac{\ell}{2}\frac{2v}{c^2 - v^2} = \ell\frac{v}{c^2}\frac{1}{1 - \left(\frac{v}{c}\right)^2}.$$

The expression for ℓ as a function of ℓ', (4.3), turns this equation into

$$t_2 - t_1 = \frac{1}{\sqrt{1 - \left(\frac{v}{c}\right)^2}}\frac{v}{c^2}\ell'.$$

For an observer in the S-frame, light reaches the left-hand mirror first and the right-hand mirror afterwards. These two events are *not* simultaneous. When ℓ' is written as $x_2' - x_1'$, the above equation and (8.1) become identical.

The *rod-through-the-house paradox* demonstrates conspicuously that *simultaneity is a relative concept*. It is the following amazing thought experiment: A rod of proper length ℓ is moving at a relativistic velocity through a house of the same proper length ℓ. In the rest frame of the house (S-frame), the rod is subject to length contraction and *fits into the house*. In the rest frame of the rod (S'-frame), the house is contracted, so the rod *does not fit*. Which statement is true? Or can both statements be true at the same time? If so, how can that be understood?

This is reminiscent of *paradoxes in philosophy*, namely statements that are both true and false. A simple example is the sentence, "I always lie". If this declaration is true, it has to be a lie as well, and there must be at least one sentence of mine that is not a lie. Then, however, the statement "I always lie" is false.

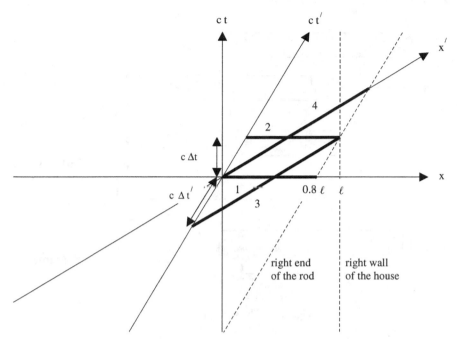

Fig. 8.3.   The rod-through-the-house paradox in a Minkowski diagram.

Let us return to the rod-through-the-house paradox. At $t = t' = 0$, the left end of the rod ($x' = 0$, i.e. the ct'-axis) and the left wall of the house ($x = 0$, i.e. the ct-axis) coincide. The rod is moving at $\beta = 0.6$, so its proper length ℓ is reduced in the S-frame to 0.8ℓ. A Minkowski diagram (Fig. 8.3) depicts the situation; the world lines of the right end of the rod and of the right wall of the house are shown as dashed lines.

The solution of the paradox is the following: Statements like "fits into the house" or "does not fit into the house" imply that *both ends* of the rod are *at the same time inside the house*, in other words simultaneously inside. However, as we know, simultaneity is a relative concept. In the S-frame, simultaneous events are on lines parallel to the x-axis. At $t = 0$, the left end of the rod and the left wall coincide (rod is line segment 1). As of this moment, the rod is fitting into the house for the time period Δt. Then the right end of the rod and the right wall coincide (rod is line segment 2), and the rod starts leaving the house. In the S'-frame, however, simultaneous

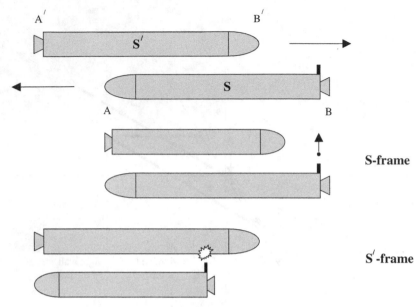

Fig. 8.4.   The star wars paradox.

events are on lines parallel to the x'-axis. When the right end of the rod reaches the right wall (rod is line segment 3), the rod is still jutting out of the left wall. During the ensuing time period $\Delta t'$, the rod is jutting out of both walls. Finally (rod is line segment 4), the left end of the rod reaches the left wall. Thus the rod *both* "fits into the house" (S-frame) *and* "does not fit" (S'-frame).

The *star wars paradox* is a closely related problem (Fig. 8.4). Two spaceships of equal proper length are passing each other at very high velocities. Spaceship S is firing a shell from its tail at right angles to its direction of propagation toward spaceship S'. Since the proper lengths are large compared with the mutual distance of the spaceships, the flight time of the shell is negligible. Let us assume that the firing, and thus the possible impact, occurs at the moment when the tip A of spaceship S passes the tail A' of spaceship S'. In the frame of spaceship S, spaceship S' is contracted, and the shot misses the target; in the frame of spaceship S', spaceship S is contracted, and the shot hits the target. *Does the shot miss or hit the target?*

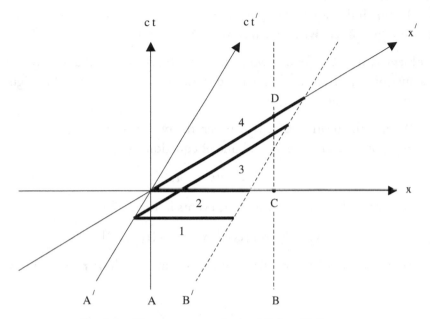

Fig. 8.5. The star wars paradox in a Minkowski diagram.

The fastest way of approaching a solution is again by means of a Minkowski diagram (Fig. 8.5). Spaceship S extends from the world line of its tip, A, to the world line of its tail, B; spaceship S' extends from A' to B'. Again, the critical feature of the Minkowski diagram is the fact that all events that are simultaneous in the S-frame are on lines parallel to the x-axis; spaceship S' is represented by the line segments 1 or 2. In the S'-frame, simultaneous events are on lines parallel to the x'-axis; spaceship S' is represented by the line segments 3 or 4.

The "moment when the tip A of spaceship S passes the tail A' of spaceship S'" is an ambiguous formulation. In the S'-frame the "moment when ..." is the x-axis, the shelling is event C; C is not a point of the line segment 2; the shot misses the target. In the S'-frame, the "moment when ..." is the x'-axis; the shot is event D; D is a point of the line segment 4; the shot hits the target. Therefore, the paradox is the result of an inaccurate use of language: There is no "moment" containing the same set of events in two inertial frames that are in relative motion to each other. *The word "moment" requires a comment in relativistic physics.*

To conclude Part I, we want to tackle four problems which illustrate the concepts of relativity explained up to this point:

Problem 1: In the S-frame, two events are happening simultaneously at a distance of 3 ly. In the S'-frame, the distance is 3.5 ly (compare the origin and point P in Fig. 7.10).

(a) What is the relative velocity between the two frames?
Solution: We take the second line of equation (6.5),

$$x' = \gamma(x - \beta ct),$$

and obtain for the distance of the events in the S'-frame

$$x'_2 - x'_1 = \gamma[(x_2 - x_1) - \beta c(t_2 - t_1)].$$

Since simultaneity is required for the S-frame, we have $t_1 = t_2$. Thus

$$\gamma = \frac{x'_2 - x'_1}{x_2 - x_1} = \frac{7}{6},$$

$$\beta = \sqrt{1 - \frac{1}{\gamma^2}} \approx 0.5.$$

(b) What is the temporal distance of the events in the S'-frame?
Solution: We take the first line of equation (6.5),

$$ct' = \gamma(ct - \beta x),$$

and obtain for the temporal distance of the events in the S'-frame

$$c(t'_2 - t'_1) = -\gamma\beta(x_2 - x_1) \approx -1.8 \, \text{ly},$$

i.e. the event at x'_2 is happening 1.8 years before the event at x'_1.

Problem 2: In the S-frame, two events are happening in one place at a temporal distance of 3 years. In the S'-frame, this distance is 3.5 years (compare the origin and point Q in Fig. 7.9).

(a) What is the relative velocity between the two frames?
Solution: We take

$$ct' = \gamma(ct - \beta x)$$

twice and subtract, considering $x_1 = x_2$,

$$c(t'_2 - t'_1) = \gamma c(t_2 - t_1).$$

Thus

$$\gamma = \frac{t'_2 - t'_1}{t_2 - t_1} = \frac{7}{6},$$

$$\beta = \sqrt{1 - \frac{1}{\gamma^2}} \approx 0.5.$$

(b) What is the spatial distance of the events in the S'-frame?
Solution: We take

$$x' = \gamma(x - \beta ct)$$

twice and subtract, considering $x_1 = x_2$,

$$x'_2 - x'_1 = -\gamma \beta c(t_2 - t_1) \approx -1.8\,\text{ly},$$

i.e. the second event is happening 1.8 ly on the left of the first event. Of course, this has to be so as the S'-frame is moving to the right at $\approx c/2$ for 3.5 years.

Problem 3: At $(ct_1 = x_0 \mid x_1 = x_0)$ and $(ct_2 = 0.5x_0 \mid x_2 = 2x_0)$, two events are happening in the S-frame. As usual, the S'-frame moves at the velocity v relative to the S-frame.

(a) What is the magnitude of v for both events to happen simultaneously in the S'-frame?
Solution: We take

$$ct' = \gamma(ct - \beta x)$$

twice and subtract, considering $t'_2 = t'_1$,

$$0 = \gamma[c(t_2 - t_1) - \beta(x_2 - x_1)].$$

Thus

$$\beta = \frac{c(t_2 - t_1)}{x_2 - x_1} = \frac{-0.5x_0}{x_0} = -0.5.$$

(b) At what times t' do these events happen in the S'-frame?
Solution: We take

$$t' = \gamma \left(t - \frac{\beta}{c} x \right),$$

and insert the numerical values of event 1,

$$t_1' = \frac{1}{\sqrt{1 - (-0.5)^2}} \left(\frac{x_0}{c} + \frac{0.5}{c} x_0 \right) \approx 1.7 \frac{x_0}{c},$$

or of event 2,

$$t_2' = \frac{1}{\sqrt{1 - (-0.5)^2}} \left(\frac{0.5 x_0}{c} + \frac{0.5}{c} 2 x_0 \right) \approx 1.7 \frac{x_0}{c},$$

and obtain the same time, as expected.

Problem 4: A spaceship leaves earth at a velocity of $\beta = 0.8$. When it is $x_0 = 6.66 \times 10^{11}$ m away from earth, earth is transmitting a radio signal toward the spaceship.

(a) How long does the radio signal travel in the earth-frame?
Solution: For such problems, a Minkowski diagram is very useful (Fig. 8.6).

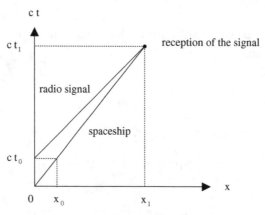

Fig. 8.6. Minkowski diagram for problem 4.

One writes down the straight-line equations for both the spaceship and the signal,

$$ct = \frac{1}{\beta}x,$$

$$ct = x + ct_0,$$

(the slope is the reciprocal (!) of the velocity normalized to c) and intersects the two lines by eliminating x,

$$\beta ct = ct - ct_0$$

$$t = \frac{t_0}{1 - \beta}.$$

The intersection is at $t_1 = 5t_0$, i.e.

$$t_1 - t_0 = 4t_0 = 4\frac{x_0}{0.8c} = 11{,}100 \text{ s}.$$

(b) How long does the radio signal travel in the spaceship-frame?
Solution: We have to Lorentz-transform the events $(ct_0|0)$ and $(ct_1|x_1)$ into the (primed) spaceship-frame. For the transformation equations to be applicable, we have to make the assumption that, when the spaceship is launched, the origins of the two coordinate frames coincide and $t = t' = 0$. The numerical values to be used in the transformation are

$$t_0 = \frac{x_0}{v} = 2{,}775 \text{ s},$$

$$t_1 = 5t_0 = 13{,}875 \text{ s},$$

$$x_1 = vt_1 = 5vt_0 = 5x_0 = 5 \times 6.66 \times 10^{11} \text{ m} = 3.33 \times 10^{12} \text{ m},$$

$$\gamma = \frac{5}{3}.$$

We take the first line of equation (6.5) twice, subtract, and obtain

$$t'_1 - t'_0 = \gamma \left[(t_1 - t_0) - \frac{\beta}{c}(x_1 - 0) \right] = 3{,}700 \text{ s}.$$

It is interesting to note that the time intervals, $t_1 - t_0$ and $t_1' - t_0'$, differ by a factor of 3 rather than 5/3, as one might believe from looking at the time dilation factor γ. An unwitting application of (2.1) would be false as we are not talking here about a *time interval elapsing at a certain location*, like on earth or in the spaceship. It is rather a *time interval between two spatially separated events* – both in the earth-frame and in the spaceship-frame. Thus, there is no alternative to Lorentz-transforming the events.

Problems for Part I

1. **Taylor expansion**
   How large is the relative error due to a first order Taylor expansion of (1.1) and (1.2), when $v = 60$ km/s is taken?

2. **Ritz's hypothesis**
   Two stars, one as massive and the other twice as massive as the sun $(m_\odot = 1.99 \times 10^{30}$ kg$)$, are at a mutual distance of 1 AU (astronomical unit $= 150 \times 10^6$ km) and are moving on circular orbits around their common center of mass.

   (a) Calculate the velocity and the period of the smaller one.
   (b) Use this result to calculate, according to Ritz's hypothesis, the shortest distance for that star to appear twice to an observer.

3. **Principle of special relativity**
   When Einstein was a youth he pondered the following problem, among others: A runner is observing himself in a mirror that he carries in front of him. Can he observe himself when he is running almost at the speed of light $(v = c - \varepsilon)$?

4. **Time dilation**
   Calculate the γ-factor of time dilation for the following velocities: 100 km/h (car); 1000 km/h (airplane); 8.0 km/s (low flying earth satellite); 250 km/s (sun around our galaxy); $\beta = 0.01$; $\beta = 0.10$; $\beta = 0.90$; $\beta = 0.99$; $\beta = 1 - \varepsilon$.

5. **Proper time I**
Calculate the proper time of a process that requires a time of 10 s and covers a distance of 1.0×10^6 km in a different inertial reference frame, e.g. ours. How do things look like when the numbers read 10 s and 3.0×10^6 km, or 10 s and 5.0×10^6 km, respectively?

6. **Proper time II**
How fast would a spaceship have to fly in order to reach, in one year's proper time, a star called Proxima Centauri (4.2 ly), the center of our galaxy (27×10^3 ly), and the neighboring Andromeda galaxy (2.4×10^6 ly), respectively?

7. **Synchronization**
Clock A remains at its place, clock B is carried around the earth (40,000 km). By how many seconds is clock B slower if it is carried in a car (100 km/h), in an airplane (1000 km/h), and in an earth satellite (8.0 km/s), respectively?

8. **Hafele–Keating experiment**
Calculate the correction due to the general theory of relativity using (13.15). Insert $g = 9.8$ m/s^2. What would be the outcome of the experiment if, instead, an earth satellite at a height of 200 km were taken?

9. **Length contraction**
In a different universe the speed of light amounts to as little as 200 km/h. A Toyota (proper length 4.00 m), a Mercedes (5.00 m), and a Cadillac (6.00 m) are on the road. The Toyota is running at 100 km/h. How fast do the Mercedes and the Cadillac have to go in order for all three cars to be equally long to an observer at rest?

10. **Light clock**
A light clock of proper length ℓ' is moving at a velocity of v parallel to the x-axis. At $t = t' = 0$, A is located at $x = x' = 0$. At this moment, a light flash is emitted from A to B. At what times t and t', respectively, does the light flash reach B?

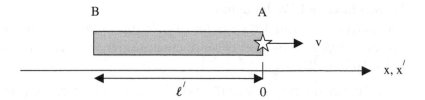

11. **Deformation of a rectangle due to length contraction**

    The sides of a rectangle are a and b. Side a and the x-axis subtend the angle θ. Draw this rectangle with $a = 6\,\text{cm}$, $b = 4\,\text{cm}$, and $\theta = 20°$. The rectangle is now moving at a velocity of v parallel to the x-axis.

    (a) Find, by construction, how length contraction with a contraction factor of $\gamma(v) = 2$ deforms the rectangle into a parallelogram.

    (b) Calculate the area of the parallelogram. (Hint: a and b are vectors)

    (c) Show that, in general, the parallelogram has an area of ab/γ.

12. **Deformation of a rectangular parallelepiped (cuboid) due to length contraction**

    (a) The vectors $\mathbf{p} = (\sqrt{3}a/2, b, -a/2)$, $\mathbf{q} = (-\sqrt{3}b/2, a, b/2)$ and $\mathbf{r} = ((a^2 + b^2)/2, 0, \sqrt{3}(a^2 + b^2)/2)$ span a rectangular parallelepiped. Give proof of this and calculate its volume.

    (b) As a result of its motion at a velocity of v parallel to the x-axis, length contraction turns it into an oblique-angled parallelepiped (spar). Calculate its volume with the triple product and show that its volume decreases by a factor of γ.

13. **Deformation of a sphere due to length contraction**

    Show that length contraction deforms a sphere in motion to an oblate rotational ellipsoid whose volume decreases by a factor of γ.

14. **Muons**

    A muon beam moves with $\beta = 0.900$ through two counters 1000 m apart. The first one counts 1000 muons per second. How many muons hit the second counter? Make your calculation both with time dilation and length contraction. $T'_{1/2}$ is $1.50\,\mu\text{s}$.

15. **Lorentz force and Coulomb force**

An electric current of 1.00 A is flowing through a copper filament (diameter 1.00 mm, number of conduction electrons per unit volume $8.4 \times 10^{28}\,\text{m}^{-3}$).

(a) Calculate the average electron velocity v, the γ-factor, the positive and negative charge densities in the S'-frame, and the charge density surplus.

(b) Calculate (in the S-frame) the Lorentz force and (in the S'-frame) the Coulomb force exerted on an elementary charge moving at a velocity of v parallel to, and at a distance of 1.00 cm from, the axis of the filament.

16. **Lorentz transformation**

Derive the inverse Lorentz transformation by solving (6.2) and (6.3) for x and then equating the results. What happens when the inverse Lorentz transformation (6.6) is inserted into the Lorentz transformation (6.5)?

17. **Events I**

The S'-frame is moving at a velocity of $\beta = 0.60$ relative to the S-frame; their origins coincide at $t = t' = 0$.

(a) Calculate and construct the space-time coordinates in S' of those four events that are located in S 4.0 ly to the left (to the right) of the origin and 2.0 years in the future (in the past).

(b) Where are all events located that pertain to the future in S, but to the past in S'?

18. **Events II**

The S''-frame is moving at a velocity of $\beta = -0.60$ relative to the S-frame; their origins coincide at $t = t'' = 0$.

(a) Calculate and construct the space-time coordinates in S'' of those four events that are located in S 4.0 ly to the left (to the right) of the origin and 2.0 years in the future (in the past).

(b) Where are all events located that pertain to the future in S, but to the past in S''?

19. **Galilei transformation and Lorentz transformation compared**
    The S_1'-, S_2'-, and S_3'-frames are moving at $\beta_1 = 0.010$, $\beta_2 = 0.100$, and $\beta_3 = 0.900$, respectively, relative to the S-frame. The origins coincide at $t = t_1' = t_2' = t_3' = 0$. Use both the Galilei and the Lorentz transformation to transform the event $(ct = 2.00\,\text{ly}\,|\,x = 5.00\,\text{ly})$ of the S-frame into the S_1'-, S_2'-, and S_3'-frames and compare. Check your results by applying the inverse Lorentz transformation, $S_3' \to S$.

20. **Simultaneity**
    In the S-frame, the event E_1 is taking place at $t_1 = 10\,\text{s}$ and $x_1 = 0\,\text{km}$, E_2 at $t_2 = 11\,\text{s}$ and $x_2 = 700{,}000\,\text{km}$, E_3 at $t_3 = 12\,\text{s}$ and $x_3 = 500{,}000\,\text{km}$. Are there any reference frames for which E_1 and E_2, and E_1 and E_3, respectively, are taking place simultaneously? If so, how fast are they moving relative to the S-frame? First visualize the problem with a Minkowski diagram and change all coordinates into light-seconds.

21. **World lines**
    What physical significance might the following eight world line diagrams have? (thick lines are material particles, thin lines are light)

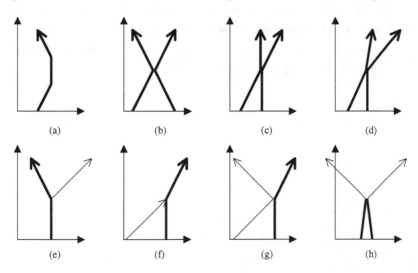

(a)   (b)   (c)   (d)

(e)   (f)   (g)   (h)

22. **Speeds faster than light**
    A star emits radiation into all directions. At a distance of 150×10^6 km, a balloon of variable shape orbits the star at a velocity of 30 km/s. The balloon is about as large as the star, thus projecting a very long umbra (core shadow) into space. The umbra is cast on a gigantic projection wall at a large distance from the star. As of which distance from the star does the umbra move at a speed faster than light? Why is it impossible to exchange information at $v > c$ by changing the shape of the balloon and thus the shape of the umbra?

23. **Spaceship and light signal**
    A spaceship (S'-frame) to the moon passes earth (S-frame) with a velocity of $\beta = 0.20$ at $t = t' = 0$. At $t = 2.0$ s, a light signal is being emitted from earth to moon (distance 400,000 km) where it gets reflected back to earth. Where and when does the light signal pass the spaceship? Solve this problem graphically and analytically for the S-frame and then transform the result into the S'-frame.

24. **Bullets and radar signal**
    Two space stations at war, A and B, are permanently at a mutual distance of 600 light seconds. Station B is shelling station A with bullets 60,000 km/s fast. How much time is left for the astronauts on station A after their early-warning radar has detected the approaching bullets? Solve the problem graphically.

PART II
Kinematics

Chapter 9

Transformation of Velocities

The problem to be tackled in this chapter is the following: The S'-frame moves at a constant velocity of $\mathbf{v} = (v, 0, 0)$ relative to the S-frame. In the S'-frame an object moves at the constant velocity

$$\mathbf{u}' = (u'_x, u'_y, u'_z) = \left(\frac{dx'}{dt'}, \frac{dy'}{dt'}, \frac{dz'}{dt'} \right).$$

What is its velocity

$$\mathbf{u} = (u_x, u_y, u_z) = \left(\frac{dx}{dt}, \frac{dy}{dt}, \frac{dz}{dt} \right)$$

in the S-frame (Fig. 9.1)?

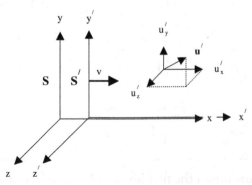

Fig. 9.1. Velocity of an object in the S'-frame.

With the inverse Lorentz transformation (6.6), one can write u_x as

$$u_x = \frac{dx}{dt} = \frac{\dfrac{dx}{dt'}}{\dfrac{dt}{dt'}} = \frac{\dfrac{d}{dt'}[\gamma(x' + \beta ct')]}{\dfrac{1}{c}\dfrac{d}{dt'}[\gamma(ct' + \beta x')]}.$$

Since v, and thus γ, are constant, γ cancels, reducing the expression to

$$u_x = \frac{\dfrac{dx'}{dt'} + \beta c\dfrac{dt'}{dt'}}{\dfrac{dt'}{dt'} + \dfrac{1}{c}\beta\dfrac{dx'}{dt'}} = \frac{u_x' + v}{1 + \dfrac{u_x' v}{c^2}}.$$

u_y is written as

$$u_y = \frac{dy}{dt} = \frac{\dfrac{dy}{dt'}}{\dfrac{dt}{dt'}} = \frac{\dfrac{dy'}{dt'}}{\gamma\left(1 + \dfrac{u_x' v}{c^2}\right)} = \frac{u_y'}{\gamma\left(1 + \dfrac{u_x' v}{c^2}\right)}.$$

A corresponding result is true for u_z. These are the equations needed to transform velocities from the S'-frame into the S-frame. Inverse equations, required for transforming velocities from the S-frame into the S'-frame, are obtained by *replacing v by $-v$ and primed variables by unprimed ones and vice versa*. The result is:

$$\boxed{\begin{array}{lll} u_x = \dfrac{u_x' + v}{1 + \dfrac{u_x' v}{c^2}} & u_y = \dfrac{u_y'}{\gamma\left(1 + \dfrac{u_x' v}{c^2}\right)} & u_z = \dfrac{u_z'}{\gamma\left(1 + \dfrac{u_x' v}{c^2}\right)} \\[4ex] u_x' = \dfrac{u_x - v}{1 - \dfrac{u_x v}{c^2}} & u_y' = \dfrac{u_y}{\gamma\left(1 - \dfrac{u_x v}{c^2}\right)} & u_z' = \dfrac{u_z}{\gamma\left(1 - \dfrac{u_x v}{c^2}\right)} \\[4ex] & \gamma = \dfrac{1}{\sqrt{1 - \left(\dfrac{v}{c}\right)^2}}. \end{array}} \qquad (9.1)$$

The first equation on the first line of (9.1) is often called the *addition theorem of velocities*. It gives an answer to the following kind of question:

A train (S'-frame) is going at the velocity v past an observer standing beside the tracks (S-frame). Inside the train an object is being thrown at the velocity u'_x in the direction of motion. What is the velocity u_x of the object for an observer beside the tracks? Two arithmetic properties of the addition theorem are striking at once:

- For $u'_x/c \ll 1$ and/or $v/c \ll 1$, the result approaches the non-relativistic limit, i.e.: $u_x = u'_x + v$.
- For $u'_x > 0$ and $v > 0$ the sum of the velocities is $u_x < u'_x + v$, a value smaller than in the non-relativistic limit.

A few numerical examples may illustrate the equations (9.1):

Example 1: The velocities are $v = u'_x = 30\,\text{km/s} = 10^{-4}c$.

This gives $u_x = 2 \times 10^{-4}c/(1 + 10^{-8}) = 60\,\text{km/s} - 0.6\,\text{mm/s}$. The deviation from the non-relativistic value, 60 km/s, is utterly negligible. The addition theorem is irrelevant for contemporary space flight.

Example 2: The velocities are $v = u'_x = 0.5c$.

This gives $u_x = 1c/(1 + 0.25) = 0.8c$. At higher velocities, e.g. those in particle accelerators, the relativistic effect is non-negligible.

Example 3: *Attempts to exceed the velocity of light, c:*

(a) The velocities are $v = c$, $u'_x = c$, and $u'_y = 0$.
This gives $u_x = 2c/(1 + 1) = c$ and $u_y = 0$. If a train is travelling at c, and inside the train an object is being cast forward at c, the object still has the velocity c, as measured by an observer on the tracks.

(b) The velocities are $v = c$, $u'_x = 0$, and $u'_y = c$.
This gives $u_x = c$ and $u_y = c/\gamma = 0$ as $\gamma \to \infty$, and therefore $u = c$. So, if the train is travelling at c, and inside the train an object is being cast at right angles to the direction of motion at the velocity c, the object has again a velocity of c, as measured by an observer beside the tracks.

From (a) and (b), we draw one of the most important conclusions of special relativity:

> The velocity of light can never be exceeded by adding velocities. The velocity, or speed, of light is the maximum speed in the universe.

Example 4: v is greater than c from the very beginning, e.g. $v = 2c$ and $u'_x = c$.

This gives $u_x = 3c/(1 + 2) = c$. Even if the inertial reference frames were able to move at superluminal velocities relative to each other, the velocity of the object would remain c in the S-frame.

This is the place to draw the reader's attention to a likely misunderstanding of the addition theorem of velocities: It may only be used *if the velocity on the left-hand side and the velocity on the right-hand side of* (9.1) *are measured in different inertial reference frames.* If two velocities, u_{x1} and u_{x2}, are measured in the same reference frame, and one wants to calculate their velocity difference, the formula is $\Delta u_x = u_{x1} - u_{x2}$, even if the velocities are relativistic. As an example, let us think of two photons travelling at $+c$ and $-c$ along a line; their velocity difference is, of course, $2c$. From the viewpoint of one photon, however, the other photon's velocity is $\pm c$ because the velocity of the observer and the second photon's velocity in the reference frame of the observer have to be added (Example 3a).

Let us pause here for a moment to remember the father of the special theory of relativity. The predictions made by Einstein, such as time dilation, length contraction, and the addition of velocities, are not only *in stark contradiction to all common sense* but, to word informally, *somewhat crazy.* But not only this. In 1905, there was no experimental justification whatsoever for special relativity. What is more, it was in blatant contradiction to Newtonian mechanics that had already been two-hundred years old at that time, which was very well proven, a downright lighthouse of exact science, and one of the intellectual cornerstones of modernity. Also, relativity was more complicated than the existing theory, an argument frequently used *against* a new theory – Einstein versus the rest of the world. And still, he

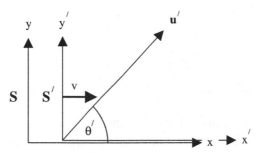

Fig. 9.2. Velocity $\mathbf{u}' = (u'_x, u'_y, 0)$ and angle θ'.

fought unwaveringly for his theory, expanding it to a general theory of relativity, utterly beyond the grasp of his contemporaries. This must be judged as an expression of supreme ingeniousness. In doing so he coined, up to the present, the conception of scientists as humans who are not being deterred from the most abstract chains of reasoning, nor from the incomprehensible conclusions inferred therefrom. He laid foundation to the notion of a scientist as a person who, armed with nothing but paper and pencil, unhinges the world as it has been known up to the present day. It is this tradition of pure theory in which modern physicists, such as cosmologists, are working today in order to treat (arithmetically, not speculatively!) such topics as the events in the very first moments of time after big bang, the characteristics of black holes, or time travels through worm holes.

We want to return to (9.1) and investigate a motion in two directions: In the S'-frame, an object has the velocity $\mathbf{u}' = (u'_x, u'_y, 0)$. The velocity vector subtends an angle θ' with the x'-axis (Fig. 9.2). What is the angle θ between $\mathbf{u}$ and the x-axis?

For the S-frame we find

$$\tan \theta = \frac{u_y}{u_x} = \frac{\dfrac{u'_y}{\gamma \left(1 + \dfrac{u'_x v}{c^2}\right)}}{\dfrac{u'_x + v}{1 + \dfrac{u'_x v}{c^2}}},$$

thus

$$\tan \theta = \frac{1}{\gamma} \frac{u'_y}{u'_x + v}. \tag{9.2}$$

Thus, with the exception of $v = 0$, $\tan \theta$ is *smaller in the S-frame than* $\tan \theta' = u'_y / u'_x$ *in the S'-frame*. For the non-relativistic limit ($\gamma \to 1$), (9.2) turns into

$$\tan \theta_{n.rel} = \frac{u'_y}{u'_x + v},$$

which is immediately obvious as $u_y = u'_y$ and $u_x = u'_x + v$.

In particular, (9.2) must not be confused with (4.5): *Here*, we are dealing with an *object in motion* in the S'-frame whose *direction of motion* **u'** subtends an angle θ' with the x'-axis. *There*, we were talking about an *object at rest* in the S'-frame whose *longitudinal direction* subtended an angle θ' with the x'-axis. *Here*, the transformation from the S'-frame to the S-frame decreases the angle, *there*, the angle increased.

The addition of velocities can also be constructed with the help of a Minkowski diagram (Fig. 9.3). Let us take example 2 with $v = u'_x = 0.5c$ and the result $u_x = 0.8c$. To demonstrate the graphical solution more generally, the launching of the object does not take place at $ct' = 0$, but after one year in the S'-frame, i.e. at $ct' = 1$ ly.

The angle between primed and unprimed axes is given by $\tan \delta = \beta = 0.5$; the unit of length on the axes of the S'-frame is, according to (7.2),

$$L = \sqrt{\frac{1 + \beta^2}{1 - \beta^2}} \approx 1.29.$$

One starts by constructing the world line of the object, traveling in the S'-frame at a velocity of

$$\frac{u'_x}{c} = \frac{\Delta x'}{\Delta ct'} = \frac{0.5 \, \text{ly}}{1 \, \text{ly}} = 0.5,$$

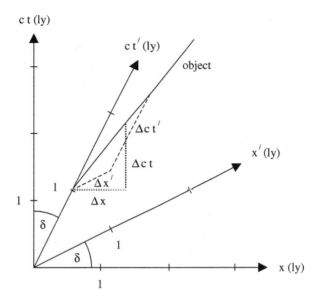

Fig. 9.3.   Addition of velocities with a Minkowski diagram.

by means of the pitch triangle in the oblique coordinates of the S'-frame
(dashed): $\Delta x' = 0.5\,\text{ly}$ and $\Delta ct' = 1\,\text{ly}$. Then another pitch triangle is
constructed on this world line in the rectangular coordinates of the S-frame
(dotted). Now we can read off: $\Delta x = 0.8\,\text{ly}$ to the right and $\Delta ct = 1\,\text{ly}$
upward; thus we obtain $u_x = 0.8\,c$.

Chapter 10*

Aberration of Light

The relativistic addition theorem of velocities can be applied to correctly explain a problem of astronomy, namely the aberration of light from a star. Let us start by assuming that an astronomer is at rest at the origin of an inertial reference frame (S-frame), i.e. not on earth, but rather on the sun, if its rotation and revolution around our galaxy, or milky way, is disregarded for the moment. He is observing a star at rest on the y'-axis of the S'-frame, and thus being in motion in the S-frame at a velocity of $\mathbf{v}$ (Fig. 10.1). Without any loss of generality, we can choose the coordinate axes such that the velocity of the star has only an x'- (x-) component, i.e. $\mathbf{v} = (v, 0, 0)$, and the position of the star is completely determined by only two coordinates, x' and y' (x and y).

The star emits light in all directions so that one ray is radiated in the direction of the astronomer at the center of the S-frame, i.e. in the direction θ', as seen from the star. To be more precise, θ' is the angle spanned by the direction of motion, $\mathbf{v}$, of the star and the direction of light emission, i.e. the wave-vector $\mathbf{k}'$, as observed in the S'-frame. The light passes the astronomer in the S-frame in the direction θ. The question is: how large is θ? To start with the calculation, we have to decompose the velocity of the light into an x'- and a y'-component: $u'_x = c \cos \theta'$ and $u'_y = c \sin \theta'$. With (9.1), we obtain for the velocity components in the S-frame

$$u_x = \frac{u'_x + v}{1 + \dfrac{u'_x v}{c^2}} = \frac{c \cos \theta' + v}{1 + \dfrac{v \cos \theta'}{c}}$$

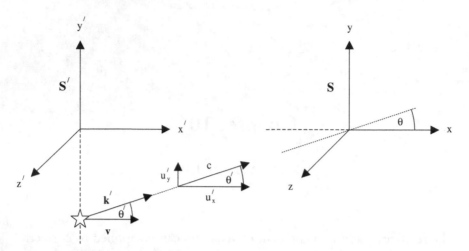

Fig. 10.1. Motion of a star in the S-frame that is at rest in the S'-frame.

$$u_y = \frac{u'_y}{\gamma\left(1 + \dfrac{u'_x v}{c^2}\right)} = \frac{c \sin \theta'}{\gamma\left(1 + \dfrac{v \cos \theta'}{c}\right)}.$$

At this point, one may want to check whether $c^2 = u_x^2 + u_y^2$ holds, as the two velocity components squared must add up to c^2 again in the S-frame as well:

$$u_x^2 + u_y^2 = \frac{c^2}{(1 + \beta \cos \theta')^2}[(\cos \theta' + \beta)^2 + (1 - \beta^2)\sin^2 \theta']$$

$$= \frac{c^2}{(1 + \beta \cos \theta')^2}[1 + 2\beta \cos \theta' + \beta^2(1 - \sin^2 \theta')] = c^2,$$

Q.E.D. Let us return to u_x and u_y, calculate

$$\sin \theta = \frac{u_y}{c} = \frac{\sin \theta'}{\gamma(1 + \beta \cos \theta')} \tag{10.1}$$

and

$$\cos \theta = \frac{u_x}{c} = \frac{\cos \theta' + \beta}{1 + \beta \cos \theta'}, \tag{10.2}$$

and find

$$\tan \theta = \frac{\sin \theta}{\cos \theta} = \frac{\sin \theta'}{\gamma(\cos \theta' + \beta)}. \tag{10.3}$$

Of course, (10.3) and (9.2) are identical. When (9.2) is extended by $1/c$, and the velocity terms are replaced by $u'_y/c \to \sin \theta'$, $u'_x/c \to \cos \theta'$, and $v/c \to \beta$, one obtains (10.3).

An "appealing" form of (10.3), more symmetric and easy to solve for the primed variable, can be elaborated with the trigonometric relationship

$$\tan \frac{\theta}{2} = \frac{\sin \theta}{1 + \cos \theta},$$

and with the help of (10.1) and (10.2):

$$\tan \frac{\theta}{2} = \frac{\dfrac{\sin \theta'}{\gamma(1 + \beta \cos \theta')}}{1 + \dfrac{\cos \theta' + \beta}{1 + \beta \cos \theta'}} = \frac{\sin \theta'}{\gamma(1 + \beta \cos \theta' + \cos \theta' + \beta)}$$

$$= \frac{\sin \theta'}{\gamma(1 + \beta)(1 + \cos \theta')} = \sqrt{\frac{1 - \beta^2}{(1 + \beta)^2}} \tan \frac{\theta'}{2}$$

$$\tan \frac{\theta}{2} = \sqrt{\frac{1 - \beta}{1 + \beta}} \tan \frac{\theta'}{2}. \tag{10.4}$$

It is clear from the square root term that for $\beta > 0$ ($\beta < 0$) *the observation angle, θ, is smaller (greater) than the emission angle, θ'.* For non-relativistic velocities ($\beta \to 0$), however, the effect is very small.

Let us now calculate the difference of these angles, $\theta' - \theta$, for very small values of β and thus for very small $\theta' - \theta$. This situation is most frequently encountered in astronomy with the exception of extragalactic astronomy. We solve (10.2) for $\cos \theta'$ and write

$$\cos \theta' = \frac{\cos \theta - \beta}{1 - \beta \cos \theta} \approx (\cos \theta - \beta)(1 + \beta \cos \theta)$$

$$\approx \cos \theta - \beta + \beta \cos^2 \theta = \cos \theta - \beta \sin^2 \theta \tag{10.5}$$

to first order in β. Then we write

$$\cos\theta' = \cos[(\theta' - \theta) + \theta] = \cos(\theta' - \theta)\cos\theta - \sin(\theta' - \theta)\sin\theta$$
$$\approx \cos\theta - (\theta' - \theta)\sin\theta \qquad (10.6)$$

to first order in $\theta' - \theta$. Comparison of (10.5) and (10.6) gives

$$\boxed{\theta' - \theta = \beta\sin\theta. \qquad (10.7)}$$

In the low velocity limit, *the aberration increases linearly with the velocity and with the sine of the angle of incidence.* For arbitrary velocities, the aberration can easily be calculated from (10.4).

Aberration, however, is of little importance for our astronomer in an almost perfect inertial reference frame as it remains constant over a very long period of time for every star. The situation is quite different for an astronomer on earth. For simplicity, let us first assume that the star under observation is at rest relative to the sun. The *earth's revolution around the sun* at $v = 30$ km/s ($\beta = 10^{-4}$) is the cause of a relative motion between the observer and the star, so the *aberration undergoes a seasonal change.*

If the star's position is on the ecliptic plane and earth is moving toward it or away from it ($\theta = 0°$ and $180°$), there is no aberration. Three months before or after these instants, $\theta = 90°$ holds, and the aberration amounts to 10^{-4} or $21''$. So the star is oscillating on the ecliptic plane at an amplitude of $21''$. Correspondingly, a star at the north pole or south pole of the ecliptic coordinate system describes a circle with a radius of $21''$, while all other stars between the ecliptic plane and the poles go in ellipses. In a quite similar manner, the *earth's rotation around its axis* is tantamount to an additional relative velocity of up to $v = 0.5$ km/s, giving rise to an additional *diurnal aberration* of up to $0.3''$.

When the star observed is no longer at rest relative to the sun, the seasonal and diurnal aberrations are not affected. The aberration resulting from the *star's motion relative to the sun* remains *unobservable* as it is at best subject to a secular variation.

In conclusion, we would like to remark that the aberration of light from a star must not be confused with the *refraction of its light in the earth's atmosphere.* The latter may effect a drastically enhanced deviation

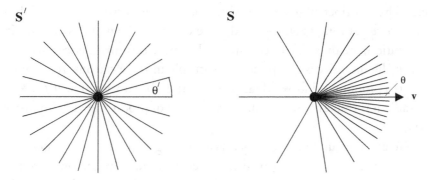

Fig. 10.2. Light rays of a source radiating isotropically in the S'-frame, moving in the S-frame at $\beta = 0.9$.

of the apparent from the actual position of the star. Nor must aberration be confused with the *annual parallax of a star*, the observation of which is restricted to near stars and never exceeds $1''$.

A further consequence of the aberration of light is a pronounced *forward radiation of fast moving radiation sources*. Radiation is called *isotropic* when its intensity is *the same into all directions*. When a radiation source (S'-frame) with an isotropic directional characteristic moves at a relativistic velocity relative to an observer (S-frame), isotropy gets lost in the observer's frame, and a *maximum of intensity* arises *in the direction of motion* (Fig. 10.2). This phenomenon is called *search light effect*. (10.4) allows us to calculate θ as a function of θ' for various values of β as a parameter:

$\theta' = 0°$	15°	30°	45°	60°	75°	90°	105°	120°	135°	150°	165°	180°
$\beta = 0.9$ $\quad\theta = 0.0°$	3.5°	7.0°	11°	15°	20°	26°	33°	43°	58°	81°	120°	180°
$\beta = 0.99$ $\quad\theta = 0.0°$	1.1°	2.2°	3.4°	4.7°	6.2°	8.1°	11°	14°	19°	30°	57°	180°
$\beta = 0.999$ $\quad\theta = 0.0°$	0.3°	0.7°	1.1°	1.5°	2.0°	2.6°	3.3°	4.4°	6.2°	9.5°	19°	180°
$\beta = 0.9999$ $\theta = 0.0°$	0.1°	0.2°	0.3°	0.5°	0.6°	0.8°	1.1°	1.4°	2.0°	3.0°	6.1°	180°

With increasing β, more and more photons that are radiated backward ($90° \leq \theta' \leq 180°$) by the source are detected in the forward direction ($0° \leq \theta \leq 90°$) by the observer, in most cases only a few degrees from the direction of motion, i.e. the source radiates highly *anisotropically*.

This relativistic effect has an important practical application in the *synchrotron* where electrons are in circular motion almost at the speed of light. The centripetal force required to confine the electrons to this path is due to a very strong magnetic field. The electrons, being permanently in acceleration, are thus able to continuously emit electromagnetic radiation. Even if the intensity is not perfectly isotropic in the electrons' rest frame, there is a pronounced forward radiation in the laboratory frame at $\beta \rightarrow 1$; the radiation is sharply focussed. This is a way of producing very intensive *X-rays*.

We can now understand how a fast moving astronaut would perceive the universe: Most photons of the universe would impinge on him at a small angle of incidence θ; almost all stars and galaxies would cluster in the neighborhood of $\theta = 0°$, i.e. in his direction of motion; the rest of the celestial sphere around him would be almost dark. It will be shown in Chapter 14 (Doppler effect) that a much higher frequency goes along with this kind of radiation, it is the life-threatening X-ray or gamma radiation.

Chapter 11

Transformation of Accelerations

The problem to be tackled in this chapter is the following: The S'-frame moves at a constant velocity of $\mathbf{v} = (v, 0, 0)$ relative to the S-frame. In the S'-frame an object moves at a constant acceleration

$$\mathbf{a}' = (a'_x, a'_y, a'_z) = \left(\frac{du'_x}{dt'}, \frac{du'_y}{dt'}, \frac{du'_z}{dt'} \right).$$

What is its acceleration

$$\mathbf{a} = (a_x, a_y, a_z) = \left(\frac{du_x}{dt}, \frac{du_y}{dt}, \frac{du_z}{dt} \right)$$

in the S-frame (Fig. 11.1)?

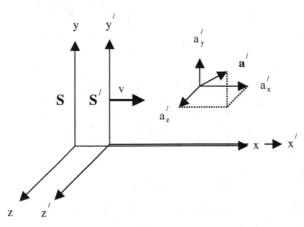

Fig. 11.1. Acceleration of an object in the S'-frame.

With the addition theorem of velocities, (9.1), and the inverse Lorentz transformation, (6.6), one obtains

$$
a_x = \frac{du_x}{dt} = \frac{\dfrac{du_x}{dt'}}{\dfrac{dt}{dt'}} = \frac{\dfrac{d}{dt'}\left[\dfrac{u'_x + v}{1 + \dfrac{u'_x v}{c^2}}\right]}{\dfrac{1}{c}\dfrac{d}{dt'}[\gamma(ct' + \beta x')]}
$$

$$
= \frac{\dfrac{du'_x}{dt'}\left(1 + \dfrac{u'_x v}{c^2}\right) - (u'_x + v)\dfrac{v}{c^2}\dfrac{du'_x}{dt'}}{\left(1 + \dfrac{u'_x v}{c^2}\right)^2 \gamma\left(\dfrac{dt'}{dt'} + \dfrac{1}{c}\beta\dfrac{dx'}{dt'}\right)}
$$

$$
= \frac{\left(1 - \dfrac{v^2}{c^2}\right) a'_x}{\left(1 + \dfrac{u'_x v}{c^2}\right)^2 \gamma\left(1 + \dfrac{u'_x v}{c^2}\right)} = \frac{1}{\gamma^3\left(1 + \dfrac{u'_x v}{c^2}\right)^3} a'_x.
$$

In addition, one obtains

$$
a_y = \frac{du_y}{dt} = \frac{\dfrac{du_y}{dt'}}{\dfrac{dt}{dt'}} = \frac{\dfrac{d}{dt'}\left[\dfrac{u'_y}{\gamma\left(1 + \dfrac{u'_x v}{c^2}\right)}\right]}{\dfrac{1}{c}\dfrac{d}{dt'}[\gamma(ct' + \beta x')]}
$$

$$
= \frac{\dfrac{du'_y}{dt'}\gamma\left(1 + \dfrac{u'_x v}{c^2}\right) - u'_y\gamma\dfrac{v}{c^2}\dfrac{du'_x}{dt'}}{\gamma^2\left(1 + \dfrac{u'_x v}{c^2}\right)^2 \gamma\left(\dfrac{dt'}{dt'} + \dfrac{1}{c}\beta\dfrac{dx'}{dt'}\right)}
$$

$$= \frac{\gamma\left(1 + \frac{u'_x v}{c^2}\right) a'_y - \gamma \frac{u'_y v}{c^2} a'_x}{\gamma^3 \left(1 + \frac{u'_x v}{c^2}\right)^2 \left(1 + \frac{u'_x v}{c^2}\right)}$$

$$= \frac{1}{\gamma^2 \left(1 + \frac{u'_x v}{c^2}\right)^2} a'_y - \frac{1}{\gamma^2 \left(1 + \frac{u'_x v}{c^2}\right)^3} \frac{u'_y v}{c^2} a'_x.$$

A corresponding result is true for a_z.

At this point one might argue that the mathematical description of accelerated motions is beyond the scope of the special theory of relativity as it is a theory of inertial reference frames only. This line of reasoning is wrong! The special theory of relativity is based rather on the assumption that the *reference frames* used for the calculations made therein (*S*-frame and *S'*-frame) are *inertial reference frames*. This is clearly the case for the above calculations. *The description of accelerated motions relative to inertial reference frames* is not subject to any limitations. We summarize:

$$a_x = \frac{1}{\gamma^3 \left(1 + \frac{u'_x v}{c^2}\right)^3} a'_x$$

$$a_y = \frac{1}{\gamma^2 \left(1 + \frac{u'_x v}{c^2}\right)^2} a'_y - \frac{1}{\gamma^2 \left(1 + \frac{u'_x v}{c^2}\right)^3} \frac{u'_y v}{c^2} a'_x$$

$$a_z = \frac{1}{\gamma^2 \left(1 + \frac{u'_x v}{c^2}\right)^2} a'_z - \frac{1}{\gamma^2 \left(1 + \frac{u'_x v}{c^2}\right)^3} \frac{u'_z v}{c^2} a'_x. \qquad (11.1)$$

The inverse equations are obtained, as usual, by interchanging primed and unprimed variables and replacing v by $-v$:

$$a'_x = \frac{1}{\gamma^3 \left(1 - \dfrac{u_x v}{c^2}\right)^3} a_x$$

$$a'_y = \frac{1}{\gamma^2 \left(1 - \dfrac{u_x v}{c^2}\right)^2} a_y + \frac{1}{\gamma^2 \left(1 - \dfrac{u_x v}{c^2}\right)^3} \frac{u_y v}{c^2} a_x$$

$$a'_z = \frac{1}{\gamma^2 \left(1 - \dfrac{u_x v}{c^2}\right)^2} a_z + \frac{1}{\gamma^2 \left(1 - \dfrac{u_x v}{c^2}\right)^3} \frac{u_z v}{c^2} a_x$$

$$\gamma = \frac{1}{\sqrt{1 - \left(\dfrac{v}{c}\right)^2}}. \tag{11.2}$$

Chapter 12

Accelerated Motion

In this chapter, we will discuss the journey of a rocket in rectilinear and constant acceleration along the x-direction. The rocket is at rest in the S'-frame; an observer left behind on earth is at rest in the S-frame.

What does travelling at a *constant acceleration* mean? A constant acceleration in the S-frame would give the astronaut a superluminal speed after a finite time. Also, a force growing beyond all limits would be required upon approaching the speed of light, as will be shown later. Therefore, we may only have a *constant acceleration in the so-called instantaneous rest frame of the rocket*. It is called the *proper acceleration* α. If it had the same magnitude as the gravitational acceleration on earth ($10\,\text{m/s}^2$), the journey would pose no health problems for the astronaut.

The instantaneous rest frame S' is an inertial reference frame for an infinitesimal time interval dt'. *At that very moment, it has the same velocity as the rocket in the S-frame, but it is not accelerated*, so the velocity of the rocket relative to S' vanishes,

$$\mathbf{u}' = (0, 0, 0). \tag{12.1}$$

Since the acceleration of the rocket in the S'-frame, i.e. its proper acceleration, is

$$\mathbf{a}' = (a'_x, 0, 0) = (\alpha, 0, 0) = \text{const}, \tag{12.2}$$

(11.1) can be used to calculate the instantaneous acceleration in the S-frame,

$$\mathbf{a} = (a_x, 0, 0) = \left(\frac{1}{\gamma^3}\alpha, 0, 0\right). \tag{12.3}$$

Since the relative velocity between S and S' always remains equal to the rocket's velocity in the S-frame,

$$v = u_x, \tag{12.4}$$

(12.3) can be written as

$$a_x = \frac{du_x}{dt} = \left(1 - \frac{u_x^2}{c^2}\right)^{3/2}\alpha. \tag{12.5}$$

During the infinitesimal time interval $dt' = dt/\gamma$, the rocket's velocity in the S-frame increases by

$$du_x = \left(1 - \frac{u_x^2}{c^2}\right)^{3/2}\alpha\,dt. \tag{12.6}$$

For consecutive infinitesimal time intervals dt, the same considerations are true, so all we have to do is integrate (12.6),

$$\int_0^{u_x} \frac{du_x}{\left(c^2 - u_x^2\right)^{3/2}} = \frac{\alpha}{c^3}\int_0^t dt.$$

The solution of the somewhat difficult integral on the left-hand side can be obtained from integral tables or with appropriate software. The result is

$$\frac{u_x}{c^2(c^2 - u_x^2)^{1/2}} = \frac{\alpha}{c^3}t$$

$$u_x^2 = \left(\frac{\alpha t}{c}\right)^2(c^2 - u_x^2)$$

$$u_x^2\left[1 + \left(\frac{\alpha t}{c}\right)^2\right] = (\alpha t)^2,$$

finally leading to the *relativistic velocity-time law*,

$$u_x = \frac{\alpha t}{\sqrt{1 + \left(\dfrac{\alpha t}{c}\right)^2}}. \tag{12.7}$$

It is always recommended that we examine a result like (12.7) with a *consistency test*, i.e. to test whether the results of the new equation are reasonable in terms of what is already known. If a result does not pass a consistency test, it is wrong; if it passes all consistency tests, it has *not* been proven correct. However, it is *more likely* to be correct, thus giving us confidence that we are on the right track. We can apply two tests here:

- For very short times, $\alpha t \ll c$ holds; so, to first order in t, the square root expression approaches unity, turning (12.7) into $u_x \approx \alpha t$. Also, the expression in parentheses in (12.5) approaches unity to first order in u_x, so we have $\alpha \approx a_x$, and $u_x \approx a_x t$ for (12.7). This is the well-known *non-relativistic velocity-time law*.
- For very long times, the second term under the root is much larger than unity, therefore $u_x \approx \alpha t/(\alpha t/c) = c$. This is a sensible result as the speed of light, c, is being approached but not exceeded.

(12.7) is by itself a differential equation which can be integrated as follows,

$$u_x = \frac{dx}{dt} = \frac{\alpha t}{\sqrt{1 + \left(\dfrac{\alpha t}{c}\right)^2}}$$

$$\int_0^x dx = \int_0^t \frac{\alpha t\, dt}{\sqrt{1 + \left(\dfrac{\alpha t}{c}\right)^2}} = \int_0^t \frac{c t\, dt}{\sqrt{\left(\dfrac{c}{\alpha}\right)^2 + t^2}}.$$

With

$$\int \frac{c t\, dt}{\sqrt{\left(\dfrac{c}{\alpha}\right)^2 + t^2}} = c\sqrt{\left(\dfrac{c}{\alpha}\right)^2 + t^2} + C,$$

we obtain

$$x = c\sqrt{\left(\frac{c}{\alpha}\right)^2 + t^2} - \frac{c^2}{\alpha},$$

and finally the *relativistic path-time law*,

$$x = \frac{c^2}{\alpha}\left(\sqrt{1 + \left(\frac{\alpha t}{c}\right)^2} - 1\right). \qquad (12.8)$$

Again, we can apply two consistency tests:

- For very short times, the second term under the square root expression is much less than unity, so the square root can be approximated by $1 + \frac{1}{2}(\frac{\alpha t}{c})^2$, thus reducing (12.8) to $x \approx \frac{1}{2}\alpha t^2$. Also, for very short times the velocity is small compared with c, so we write $\alpha \approx a_x$ with the help of (12.5), and find $x \approx \frac{1}{2}a_x t^2$, the *non-relativistic path-time law*.

- For very long times, the second term under the square root is much larger than the first and thus the largest one inside the parentheses, leaving us with $x \approx ct$. This expression gives the distance covered for a velocity close to c. But, since the velocity does actually approach c as time passes, the approximation is, indeed, sensible for long times.

Finally, we can insert the velocity-time law, (12.7), into the transformation law for accelerations, (12.5),

$$a_x = \left[1 - \frac{\left(\frac{\alpha t}{c}\right)^2}{1 + \left(\frac{\alpha t}{c}\right)^2}\right]^{3/2}\alpha,$$

and arrive at the *relativistic acceleration-time law*,

$$a_x = \frac{1}{\left[1 + \left(\frac{\alpha t}{c}\right)^2\right]^{3/2}}\alpha. \qquad (12.9)$$

A consistency test

- for very short times gives $a_x \approx \alpha$, i.e. the non-relativistic result;
- for very long times leads to $a_x \approx \frac{c^3}{\alpha^2} \frac{1}{t^3} \to 0$. As expected, the velocity increase comes eventually to a halt; otherwise the speed of light would be surpassed at some point in the journey. The fact that the speed of light can never be exceeded is the unifying thread through all of the special theory of relativity, starting with the γ-factor (2.3) which would become undefined for $v = c$.

A sample calculation with $\alpha = 10 \text{ m/s}^2$ shows us the *difference between a relativistic and a non-relativistic approach* to our space journey, and demonstrates that *after a certain time a relativistic treatment is indispensable* (Figs. 12.1, 12.2, and 12.3).

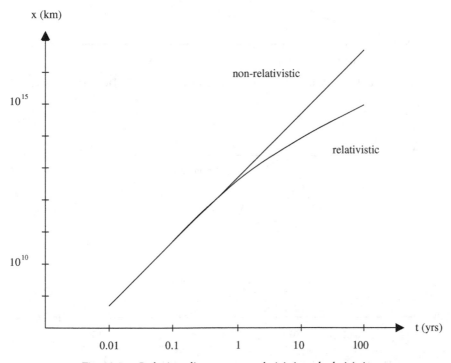

Fig. 12.1.   Path-time diagram: non-relativistic and relativistic.

t (yrs)		0.01	0.1	1	10	100
x (km)	non-relativistic	4.97×10^8	4.97×10^{10}	4.97×10^{12}	4.97×10^{14}	4.97×10^{16}
	relativistic	4.97×10^8	4.96×10^{10}	4.06×10^{12}	8.60×10^{13}	9.37×10^{14}
v (km/s)	non-relativistic	3.15×10^3	3.15×10^4	3.15×10^5	3.15×10^6	3.15×10^7
	relativistic	3.15×10^3	3.14×10^4	2.17×10^5	2.99×10^5	3.00×10^5
a_x (m/s^2)	non-relativistic	10.0	10.0	10.0	10.0	10.0
	relativistic	10.0	9.84	3.27	0.01	0.00

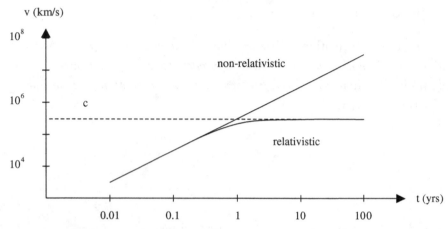

Fig. 12.2. Velocity-time diagram: non-relativistic and relativistic.

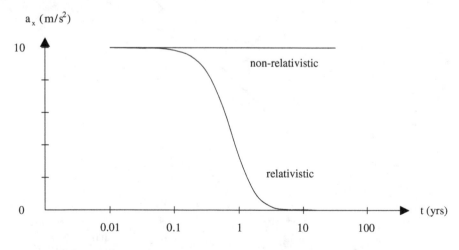

Fig. 12.3. Acceleration-time diagram: non-relativistic and relativistic.

The table of values and Fig. 12.1 show that after about one year *the difference widens between the relativistic and the non-relativistic calculation* of the distance covered by the rocket. After about a decade, the non-relativistic calculation overestimates the path by a whole order of magnitude. For short times, the non-relativistic approximation, $x \propto t^2$, is applicable, for long times, the relativistic approximation, $x \propto t$, holds. In a diagram with two logarithmic axes, $x \propto t^2$ is represented by a straight line with a slope of two, and $x \propto t$ is a line with a slope of one. These slopes are obvious in the diagram, taking into account that two powers of ten on the vertical axis are as large as one power of ten on the horizontal axis.

Figure 12.2 shows that the velocity, calculated non-relativistically, would surpass c, i.e. the *universal speed limit*, after less than a year. The relativistic velocity, however, approaches asymptotically the value of $c = 3 \times 10^5$ km/s. In the non-relativistic calculation, $v \propto t$ holds, in a double-logarithmic diagram the slope is thus one.

In a relativistic calculation, the rocket's acceleration (Fig. 12.3) is visibly falling behind its starting value of 10 m/s^2 after about one month and then approaches zero rapidly. This behavior does not surprise us because once the speed of light has almost been reached, *only infinitesimal acceleration is possible.*

Finally, the *proper time τ of the rocket* shall be considered. When t' is replaced by τ, the differential form of time dilation, (2.1), is given by

$$d\tau = dt \sqrt{1 - \left(\frac{u_x(t)}{c}\right)^2}. \tag{12.10}$$

$d\tau$ is an infinitesimal increment of proper time of the rocket (S'-frame), dt a time increment of an observer on earth (S-frame), and $u_x(t)$ the rocket's velocity relative to earth. We now insert the velocity-time law, (12.7),

$$d\tau = dt \left[1 - \frac{\left(\frac{\alpha t}{c}\right)^2}{1 + \left(\frac{\alpha t}{c}\right)^2} \right]^{1/2}$$

$$\int_0^\tau d\tau = \int_0^t \frac{dt}{\sqrt{1 + \left(\frac{\alpha t}{c}\right)^2}}$$

$$\tau = \frac{c}{\alpha} \int_0^t \frac{dt}{\sqrt{\left(\frac{c}{\alpha}\right)^2 + t^2}} = \frac{c}{\alpha} \ln\left(\frac{\alpha t}{c} + \sqrt{\left(\frac{\alpha t}{c}\right)^2 + 1}\right)\Bigg|_0^t$$

$$\tau = \frac{c}{\alpha} \ln\left(\frac{\alpha t}{c} + \sqrt{1 + \left(\frac{\alpha t}{c}\right)^2}\right). \qquad (12.11)$$

This equation expresses the proper time of the rocket as a function of the time elapsed on earth. A table of values for $\alpha = 10\,\mathrm{m/s^2}$ shows that τ grows much more slowly than t:

t (yrs)	0.01	0.1	1	10	100	1000
τ (yrs)	0.01	0.10	0.87	2.90	5.09	7.28

This circumstance can, but only theoretically, be used to make astronauts go on very long journeys, so far and so long, as seen by an observer on earth, that the astronauts would no longer recognize, or even find, earth after having spent some decades of proper time in their spaceship. These intriguing ideas will be elaborated in Chapter 15 (twin paradox) and Chapter 21 (motion of a rocket).

Let us do a consistency test for very short times. A Taylor expansion of the square root of (12.11) gives $1 + \frac{1}{2}(\frac{\alpha t}{c})^2 - \cdots$, the ln-term turns into

$$\ln\left[1 + \frac{\alpha t}{c} + \frac{1}{2}\left(\frac{\alpha t}{c}\right)^2 - \cdots\right].$$

As usual, the second term inside the brackets is much less than unity and the third term is again much less than the previous term and therefore negligible in a first order expansion. Yet another first order Taylor expansion gives

$$\ln\left(1 + \frac{\alpha t}{c}\right) = \frac{\alpha t}{c} - \cdots,$$

which leads us to the final result,

$$\tau \approx \frac{c}{\alpha} \frac{\alpha t}{c} = t.$$

For very short times, the time elapsed on earth, t, and the rocket's proper time, τ, are comparable.

The inverse function of (12.11), $t = f(\tau)$, may also be useful. To obtain it, we take the antilogarithm of (12.11),

$$\exp\left(\frac{\alpha\tau}{c}\right) = \frac{\alpha t}{c} + \sqrt{1 + \left(\frac{\alpha t}{c}\right)^2},$$

isolate the square root on the right-hand side and square,

$$\exp\left(\frac{2\alpha\tau}{c}\right) - 2\frac{\alpha t}{c}\exp\left(\frac{\alpha\tau}{c}\right) + \left(\frac{\alpha t}{c}\right)^2 = 1 + \left(\frac{\alpha t}{c}\right)^2,$$

multiply by $\exp\left(-\frac{\alpha\tau}{c}\right)$ and rearrange,

$$\exp\left(\frac{\alpha\tau}{c}\right) - \exp\left(-\frac{\alpha\tau}{c}\right) = 2\frac{\alpha t}{c}.$$

The left-hand side is twice the hyperbolic sine function with the argument $\alpha\tau/c$, so

$$t = \frac{c}{\alpha}\sinh\left(\frac{\alpha\tau}{c}\right). \tag{12.12}$$

Insertion of (12.12) into the path-time law, (12.8), yields

$$x = \frac{c^2}{\alpha}\left(\sqrt{1 + \sinh^2\left(\frac{\alpha\tau}{c}\right)} - 1\right),$$

i.e. the *path-proper time law*,

$$x = \frac{c^2}{\alpha}\left[\cosh\left(\frac{\alpha\tau}{c}\right) - 1\right]. \tag{12.13}$$

Insertion of (12.12) into the velocity-time law, (12.7), yields

$$u_x = \frac{c \sinh\left(\dfrac{\alpha\tau}{c}\right)}{\sqrt{1 + \sinh^2\left(\dfrac{\alpha\tau}{c}\right)}} = \frac{c \sinh\left(\dfrac{\alpha\tau}{c}\right)}{\cosh\left(\dfrac{\alpha\tau}{c}\right)},$$

i.e. the *velocity-proper time law*,

$$u_x = c \tanh\left(\frac{\alpha\tau}{c}\right). \tag{12.14}$$

Chapter 13*

The Rocket-Rope Paradox (Bell's Paradox)

This chapter presents an extensive, although elementary, mathematical development which, starting with the geometry of a Minkowski diagram, will carry us to one of the best-known results of the general theory of relativity.

Two rockets, A and B, are being launched in the S-frame at $(ct = 0 \mid x = 0)$ and $(ct = 0 \mid x = \ell)$, and are subject to the same constant proper acceleration α. A rope of proper length ℓ is stretched taut between the rockets. The rope will snap. But why?

For rocket A, equation (12.8) gives us the relativistic path-time law, or world line, in the S-frame,

$$x = \frac{c^2}{\alpha} \left(\sqrt{1 + \left(\frac{\alpha t}{c}\right)^2} - 1 \right). \tag{13.1}$$

Correspondingly, the world line of rocket B is given by

$$x = \frac{c^2}{\alpha} \left(\sqrt{1 + \left(\frac{\alpha t}{c}\right)^2} - 1 \right) + \ell. \tag{13.2}$$

Both rockets have the same constant mutual distance ℓ in the S-frame. Let us now have a look at the length of the rope: In the moving reference frame of the rope, its proper length ℓ remains constant. However, in the terrestrial S-frame, the length is contracted to $\ell/\gamma < \ell$, so the rope snaps. But how can we explain the snapping in the S'-frame of rocket A?

Let us first rearrange the world lines of the two rockets in such a way that their *hyperbolic character* becomes apparent; for rocket A

$$\left(\frac{\alpha}{c^2}x + 1\right)^2 = \frac{\alpha^2}{c^4}(ct)^2 + 1$$

$$\frac{\left(x + \frac{c^2}{\alpha}\right)^2}{\left(\frac{c^2}{\alpha}\right)^2} - \frac{(ct)^2}{\left(\frac{c^2}{\alpha}\right)^2} = 1, \tag{13.3}$$

and for rocket B

$$\frac{\left(x - \ell + \frac{c^2}{\alpha}\right)^2}{\left(\frac{c^2}{\alpha}\right)^2} - \frac{(ct)^2}{\left(\frac{c^2}{\alpha}\right)^2} = 1. \tag{13.4}$$

Obviously, the vertices of the hyperbolae are shifted to the left by c^2/α, (13.3), and to the right by $\ell - c^2/\alpha$, (13.4), respectively. Figure 13.1 shows the first quadrant only.

After a time t_A, rocket A has arrived at point $P(ct_A \mid x_A)$ with the coordinates

$$P\left(ct_A \left|\frac{c^2}{\alpha}\left(\sqrt{1 + \left(\frac{\alpha t_A}{c}\right)^2} - 1\right)\right.\right).$$

It is now helpful to express these coordinates by means of (12.7) with the velocity at P as a parameter,

$$\beta_A = \frac{\dfrac{\alpha t_A}{c}}{\sqrt{1 + \left(\dfrac{\alpha t_A}{c}\right)^2}}.$$

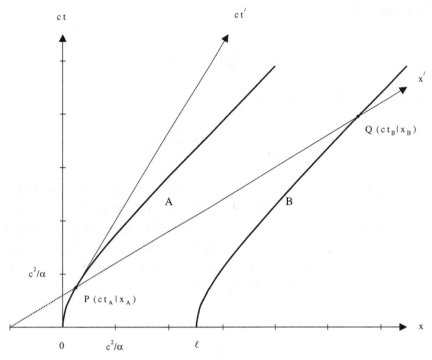

Fig. 13.1. Hyperbolic world lines of two rockets, A and B.

To do so, we solve for t_A,

$$\beta_A^2 = \left(\frac{\alpha^2}{c^2} - \beta_A^2 \frac{\alpha^2}{c^2} \right) t_A^2$$

$$t_A = \frac{c}{\alpha} \beta_A \gamma_A,$$

and insert t_A into x_A,

$$x_A = \frac{c^2}{\alpha} \left(\sqrt{1 + \beta_A^2 \gamma_A^2} - 1 \right).$$

With

$$1 + \beta_A^2 \gamma_A^2 = 1 + \frac{\beta_A^2}{1 - \beta_A^2} = \frac{1}{1 - \beta_A^2} = \gamma_A^2, \qquad (13.5)$$

we obtain

$$x_A = \frac{c^2}{\alpha}(\gamma_A - 1),$$

so the coordinates of $P(ct_A | x_A)$ are

$$P\left(\frac{c^2}{\alpha}\beta_A\gamma_A \,\middle|\, \frac{c^2}{\alpha}(\gamma_A - 1)\right). \tag{13.6}$$

We now span an S'-coordinate frame with the origin in P and oblique-angled axes, ct' and x', the tilts of which are given by the instantaneous velocity β_A of rocket A at P, according to (7.1). The origin of this S'-frame is being carried along with rocket A; so the tilts of the axes toward the vertical and the horizontal, respectively, increase with time. In the S-frame, the x'-axis is given by the straight line equation

$$ct - ct_A = \beta_A(x - x_A)$$

$$ct - \frac{c^2}{\alpha}\beta_A\gamma_A = \beta_A\left[x - \frac{c^2}{\alpha}(\gamma_A - 1)\right],$$

or, after rearranging terms,

$$ct = \beta_A\left(x + \frac{c^2}{\alpha}\right). \tag{13.7}$$

The x'-axis can be constructed by drawing a straight line through a point with the coordinates $(ct = 0 | x = -c^2/\alpha)$ and the instantaneous position of the rocket at point P (dotted in Fig. 13.1). The x'-axis is so important because all events on it, e.g. $Q(ct_B | x_B)$, occur simultaneously in the S'-frame. One can immediately see that point Q, the intersection of the x'-axis with the world line of rocket B, is rapidly moving to the right.

We now calculate x_B, the x-coordinate of Q, by computing the intersection of the x'-axis and the world line of rocket B, i.e. by inserting (13.7) into (13.2),

$$x_B = \frac{c^2}{\alpha}\left(\sqrt{1 + \frac{\alpha^2}{c^4}\beta_A^2\left(x_B + \frac{c^2}{\alpha}\right)^2} - 1\right) + \ell.$$

We solve this for x_B,

$$\left[(x_B - \ell)\frac{\alpha}{c^2} + 1\right]^2 = 1 + \frac{\alpha^2}{c^4}\beta_A^2\left(x_B + \frac{c^2}{\alpha}\right)^2$$

$$(x_B - \ell)^2 + \frac{2c^2}{\alpha}(x_B - \ell) = \beta_A^2\left(x_B^2 + 2x_B\frac{c^2}{\alpha} + \frac{c^4}{\alpha^2}\right)$$

$$\frac{x_B^2}{\gamma_A^2} + x_B\left(\frac{2c^2}{\alpha\gamma_A^2} - 2\ell\right) - \frac{2c^2}{\alpha}\ell - \beta_A^2\frac{c^4}{\alpha^2} + \ell^2 = 0$$

$$x_B = \frac{1}{2}\left[\left(2\gamma_A^2\ell - \frac{2c^2}{\alpha}\right)\right.$$

$$\left.\pm\sqrt{4\gamma_A^4\ell^2 - 8\gamma_A^2\ell\frac{c^2}{\alpha} + \frac{4c^4}{\alpha^2} + 8\gamma_A^2\frac{c^2}{\alpha}\ell + 4\gamma_A^2\beta_A^2\frac{c^4}{\alpha^2} - 4\gamma_A^2\ell^2}\right]$$

$$x_B = \gamma_A^2\ell - \frac{c^2}{\alpha} \pm \sqrt{\frac{c^4}{\alpha^2}\gamma_A^2 + \gamma_A^4\ell^2\beta_A^2}, \tag{13.8}$$

having made use of $1 + \beta_A^2\gamma_A^2 = \gamma_A^2$, (13.5), and

$$1 - \frac{1}{\gamma_A^2} = 1 - (1 - \beta_A^2) = \beta_A^2.$$

The second solution of this quadratic equation represents the intersection of the x'-axis with the second branch of the hyperbola, not included in Fig. 13.1. We can now write the coordinates of $Q(ct_B \mid x_B)$ with (13.7) and (13.8) as

$$Q\left(\beta_A\gamma_A^2\ell + \beta_A\sqrt{\frac{c^4}{\alpha^2}\gamma_A^2 + \beta_A^2\gamma_A^4\ell^2} \,\middle|\, \gamma_A^2\ell - \frac{c^2}{\alpha} + \sqrt{\frac{c^4}{\alpha^2}\gamma_A^2 + \beta_A^2\gamma_A^4\ell^2}\right).$$

$$\tag{13.9}$$

With the Pythagorean theorem, as well as (13.9) and (13.6), the length squared of the line segment PQ in the S-frame is

$$(ct_B - ct_A)^2 + (x_B - x_A)^2$$

$$= \left(\beta_A \gamma_A^2 \ell + \beta_A \sqrt{\frac{c^4}{\alpha^2} \gamma_A^2 + \beta_A^2 \gamma_A^4 \ell^2} - \frac{c^2}{\alpha} \beta_A \gamma_A \right)^2$$

$$+ \left(\gamma_A^2 \ell + \sqrt{\frac{c^4}{\alpha^2} \gamma_A^2 + \beta_A^2 \gamma_A^4 \ell^2} - \frac{c^2}{\alpha} \gamma_A \right)^2$$

$$= (\beta_A^2 + 1) \left[\gamma_A^2 \ell + \frac{c^2}{\alpha} \gamma_A \left(\sqrt{1 + \frac{\alpha^2}{c^4} \beta_A^2 \gamma_A^2 \ell^2} - 1 \right) \right]^2 .$$

Thus, the line segment PQ has a length of

$$\sqrt{\frac{1 + \beta_A^2}{1 - \beta_A^2}} \left[\gamma_A \ell + \frac{c^2}{\alpha} \left(\sqrt{1 + \frac{\alpha^2}{c^4} \beta_A^2 \gamma_A^2 \ell^2} - 1 \right) \right] .$$

Since, according to (7.2), the length unit on the S'-axes amounts to $\sqrt{\frac{1+\beta_A^2}{1-\beta_A^2}}$, the length of the line segment PQ in the S'-frame, i.e. the position of Q on the x'-axis, is reduced to

$$x' = \gamma_A \ell + \frac{c^2}{\alpha} \left(\sqrt{1 + \frac{\alpha^2}{c^4} \beta_A^2 \gamma_A^2 \ell^2} - 1 \right) . \tag{13.10}$$

This line segment is clearly longer than ℓ, thus the rope has to snap in the S'-frame as well. The mutual distance of the rockets increases, because *rocket B is not only undergoing an accelerated motion in the S-frame, but also in the S'-frame of rocket A!* This behavior is immediately obvious from the Minkowski diagram (Fig. 13.1) where world line B has a smaller slope in Q, i.e. a higher velocity, than world line A in P. This is true for both the S-frame and the S'-frame. But only in the S'-frame, B is in accelerated motion away from A; in the S-frame, this is certainly not the case as simultaneous events are always located on parallels to the x-axis. That the mutual distance of the rockets is *both accelerated and not accelerated, depending on the reference frame*, is again a result of the fact that *simultaneity is not an absolute but a relative notion.*

It is also interesting to see how the mutual distance of the rockets keeps growing in the S'-frame: For $\beta_A \to 1$, $\gamma_A \to \infty$ holds, so (13.10) turns into

$$x' \approx \gamma_A \ell + \gamma_A \ell = 2\gamma_A \ell. \tag{13.11}$$

The mutual distance becomes infinite!

Now the problem has been solved, but *just for fun* we continue our calculations a little further to answer the question of how much proper time τ_A has elapsed in rocket A when it arrives at P, and how much proper time τ_B is required for rocket B to reach Q. To do so, we take t_A from (13.6) and insert it into (12.11),

$$\tau_A = \frac{c}{\alpha} \ln\left(\beta_A \gamma_A + \sqrt{1 + \beta_A^2 \gamma_A^2}\right), \tag{13.12}$$

then take t_B from (13.9) and insert it into (12.11),

$$\tau_B = \frac{c}{\alpha} \ln\left(D + \sqrt{1 + D^2}\right), \tag{13.13}$$

with D being

$$D = \frac{\alpha}{c^2} \beta_A \gamma_A^2 \ell + \frac{\alpha}{c^2} \beta_A \sqrt{\frac{c^4}{\alpha^2} \gamma_A^2 + \beta_A^2 \gamma_A^4 \ell^2}$$

$$= \frac{\alpha}{c^2} \beta_A \gamma_A^2 \ell + \beta_A \gamma_A \sqrt{1 + \frac{\alpha^2 \ell^2}{c^4} \beta_A^2 \gamma_A^2}.$$

Let us examine the short-time limits ($t_A \to 0$). From (12.7) we find, to first order in t_A, the relation $u_A \approx \alpha t_A$, or

$$\beta_A \approx \frac{\alpha t_A}{c} \ll 1.$$

Then we expand (13.12) to first order in β_A, note that $\gamma_A \to 1$, replace β_A by $\alpha t_A / c$, and obtain

$$\tau_A \approx \frac{c}{\alpha} \ln(1 + \beta_A) = \frac{c}{\alpha} \ln\left(1 + \frac{\alpha t_A}{c}\right) \approx t_A.$$

Similarly, we expand (13.13),

$$D \approx \frac{\alpha^2 \ell}{c^3} t_A + \frac{\alpha}{c} t_A,$$

so that to first order in t_A,

$$\tau_B \approx \frac{c}{\alpha} \ln\left[\frac{\alpha}{c}\left(1 + \frac{\alpha\ell}{c^2}\right)t_A + 1\right] \approx \left(1 + \frac{\alpha\ell}{c^2}\right)t_A.$$

This gives

$$\frac{\tau_B}{\tau_A} \approx 1 + \frac{\alpha\ell}{c^2}. \tag{13.14}$$

This result is confirmed by the *general theory of relativity!* For it postulates that $-\alpha$, the inertial acceleration that A and B are continuously being exposed to, is identical to a gravitational acceleration in a homogeneous gravitational field, $-g$. Therefore, we take (13.14), replace α by g, the distance ℓ by a height difference h, and write

$$\frac{\tau_B}{\tau_A} \approx 1 + \frac{gh}{c^2}. \tag{13.15}$$

This is a formula found in many elementary treatises. It says that a clock at B, the front or the higher observer, is faster by a factor of $1 + gh/c^2$ than a clock at A, the rear or the lower observer. This type of time dilation has already been mentioned in connection with the Hafele–Keating experiment (Chapter 3).

Chapter 14

Doppler Effect

In non-relativistic physics, the *frequency shift of sound* can be calculated when the source of the sound and the receiver are in relative motion to each other. Since sound, as opposed to light, needs a medium, i.e. a material carrier such as air or water, and propagates relative to that medium with the speed of sound, v_s, one distinguishes three cases:

Case 1: *The receiver R is at rest relative to the medium, and the source S is moving away from the receiver at the velocity v (Fig. 14.1). While the source is moving from $S^{(1)}$ to $S^{(2)}$, it is undergoing one full oscillation, thus being in phase at $S^{(1)}$ and $S^{(2)}$. In the meantime, the excitation emitted at $S^{(1)}$ has turned into a spherical*

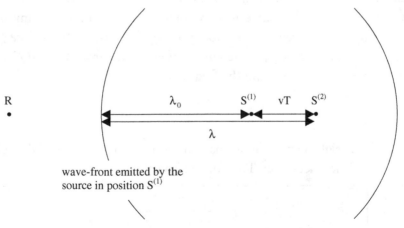

wave-front emitted by the
source in position $S^{(1)}$

Fig. 14.1. Doppler effect: Source in motion relative to the receiver.

wave propagating through space. As a result, *the wavelength in the direction of the receiver is stretched* from λ_0 to $\lambda = \lambda_0 + vT$, with T being the period.

The frequency observed by the receiver R changes from $\nu_0 = v_s/\lambda_0$ to $\nu = v_s/\lambda$, and the frequency ratio is

$$\frac{\nu}{\nu_0} = \frac{\lambda_0}{\lambda} = \frac{\lambda_0}{\lambda_0 + vT}.$$

The right-hand side of the above equation will not change when the calculation is redone relativistically as λ_0 and vT are being measured in the frame of the source. Therefore, the addition theorem of velocities is not applicable in the denominator, and the length contraction factors appearing in the frame of the receiver cancel out. The non-relativistic result reads

$$\frac{\nu}{\nu_0} = \frac{1}{1 + \dfrac{v}{v_s}}. \tag{14.1}$$

ν_0 is the frequency of the source, and ν is the frequency arriving at the receiver. The velocity v is positive when source and receiver are moving apart in which case the frequency is lowered.

Case 2: *The source is at rest relative to the medium, and the receiver is moving away from the source at the velocity v. The sound waves impinge upon the receiver at a velocity of $v_s - v$, instead of v_s. The frequency observed by the receiver is thus reduced* from $\nu_0 = v_s/\lambda_0$ to $\nu = (v_s - v)/\lambda_0$, and the frequency ratio is

$$\frac{\nu}{\nu_0} = \frac{v_s - v}{v_s}.$$

The right-hand side of this equation will not change in the relativistic calculation. The non-relativistic result is

$$\frac{\nu}{\nu_0} = 1 - \frac{v}{v_s}. \tag{14.2}$$

ν, ν_0 and v, including its sign, are the same as above.

Case 3: *Both the source and the receiver are in motion* relative to the medium. A treatment of this case is dispensed with as it is of no importance for what follows.

The two results, (14.1) and (14.2), are obviously different. The propagation of sound does not only depend on the relative velocity between source and receiver, but also on the movements of both source and receiver relative to the carrier medium. For $v/v_s \ll 1$, however, $(1 + \frac{v}{v_s})^{-1} \approx 1 - \frac{v}{v_s}$ holds; so (14.1) and (14.2) become equal.

Let us now consider the case where the speed of sound, v_s, is replaced by the speed of light, c. v is a very high, i.e. relativistic, speed. Cases 1 and 2, calculated non-relativistically, have to be subjected to the following corrections:

Case 1: In a frame where the *receiver is at rest*, the source is subject to time dilation. The source is emitting fewer wave fronts per second (time of the receiver), the frequency of emission is reduced from v_0 to $v_0\sqrt{1 - \beta^2}$. So we obtain from (14.1)

$$\frac{v}{v_0\sqrt{1 - \beta^2}} = \frac{1}{1 + \beta}$$

$$\frac{v}{v_0} = \sqrt{\frac{1 - \beta^2}{(1 + \beta)^2}}$$

$$\frac{v}{v_0} = \sqrt{\frac{1 - \beta}{1 + \beta}}. \tag{14.3}$$

Case 2: In a frame where the *source is at rest*, the receiver is subject to time dilation. The receiver is receiving fewer wave fronts per second (time of the source), the frequency of reception is reduced from v to $v\sqrt{1 - \beta^2}$. So we obtain from (14.2)

$$\frac{v\sqrt{1 - \beta^2}}{v_0} = 1 - \beta$$

$$\frac{v}{v_0} = \sqrt{\frac{(1 - \beta)^2}{1 - \beta^2}} = \sqrt{\frac{1 - \beta}{1 + \beta}}. \tag{14.4}$$

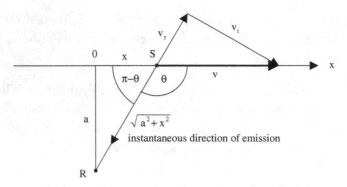

Fig. 14.2.   Source S passing the receiver R at a distance a.

The frequency ratio is the same in both cases; they become *indistin-guishable.* This has to be so because light does not require a medium, or supporting material, which means there cannot be a relative velocity between source and medium or receiver and medium, respectively.

The Doppler effect may now be generalized for the case of *a source S passing the receiver R at a distance a* (Fig. 14.2). Its velocity v has to be decomposed into a *radial component*, v_r, and a *tangential component*, v_t. From a similarity consideration of the two triangles, we see that

$$v_r = \frac{x}{\sqrt{a^2 + x^2}} v.$$

In (14.1), v has to be replaced by v_r. In the relativistic generalization of (14.1) into (14.3), the factor $\sqrt{1 - \beta^2}$ remains unaltered. For time dilation is a function of the velocity v, not only of its radial component, v_r, as has already been pointed out in Chapter 2. So the derivation of (14.3) changes to

$$\frac{\nu}{\nu_0} = \frac{\sqrt{1 - \beta^2}}{1 + \dfrac{v_r}{c}} = \frac{\sqrt{1 - \beta^2}}{1 + \dfrac{x}{\sqrt{a^2 + x^2}} \beta} = \frac{1}{\gamma(1 - \beta \cos \theta)}. \tag{14.5}$$

θ is the angle between the relative motion and the instantaneous direction of emission in the receiver's frame.

From the generalized Doppler formula, (14.5), we can easily calculate *two important limiting cases*:

- For large x, the factor $x/\sqrt{a^2 + x^2}$ approaches unity (or $\theta \to 180°$), so (14.3) is found again. The transverse velocity, v_t, is small, i.e. source and receiver are moving apart almost on a straight line which was the supposition for (14.3). This limiting case is called the *longitudinal Doppler effect*.
- For $x = 0$ (or $\theta = 90°$), when the distance between source and receiver is at a minimum, (14.5) is simplified to

$$\frac{v}{v_0} = \sqrt{1 - \beta^2}. \tag{14.6}$$

Since the radial velocity, v_r, vanishes, this limiting case is called the *transverse Doppler effect*. The frequency shift is exclusively the result of time dilation, so the effect is purely relativistic.

It is interesting to compare the longitudinal and the transverse Doppler effects: A Taylor expansion of (14.3), up to the first order in β, gives

$$\frac{v}{v_0} \approx \left(1 - \frac{1}{2}\beta\right)\left(1 - \frac{1}{2}\beta\right) \approx 1 - \beta,$$

and of (14.6), up to the second order in β, gives

$$\frac{v}{v_0} \approx 1 - \frac{1}{2}\beta^2.$$

The longitudinal (normal) Doppler effect depends linearly on velocity, the transverse one quadratically. Therefore, it is only observable at distinctly higher velocities. The frequency ratios associated with fly-bys at $\beta = 0.1$ and $\beta = 0.9$ are depicted in Fig. 14.3.

We are now going to re-derive the relativistic Doppler equation with the help of a Minkowski diagram: From the origin of the S-frame, light signals are being emitted into the x-direction at intervals of length T. Their world lines are dashed (Fig. 14.4). The event A is the absorption of the light signal emitted at $t = T$ in the S'-frame.

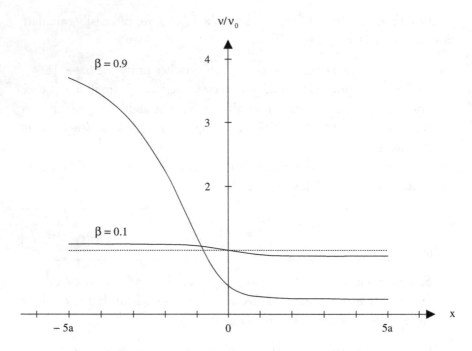

Fig. 14.3.   Frequency shift during the fly-by of a source at the distance *a*.

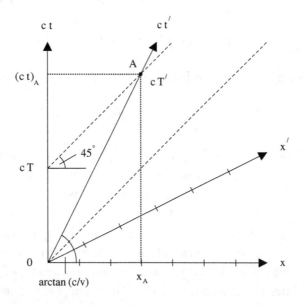

Fig. 14.4.   Derivation of (14.3) with help of a Minkowski diagram.

Let us first calculate the coordinates of A in the S-frame. To do so, we write down the equation of the world line of the light signal going through A (slope $= 1$),

$$ct = cT + x,$$

the equation of the ct'-axis (slope $= c/v$),

$$ct = \frac{c}{v}x = \frac{1}{\beta}x,$$

and equate both expressions,

$$cT + x = \frac{1}{\beta}x.$$

This gives us the x-coordinate of A,

$$x_A \left(\frac{1}{\beta} - 1 \right) = cT$$

$$x_A = \frac{\beta cT}{1 - \beta},$$

and the ct-coordinate of A,

$$(ct)_A = \frac{1}{\beta}x_A = \frac{cT}{1 - \beta}.$$

The S'-frame receives these light signals in intervals of T'. But we must not forget that, according to (7.2), length units on the S'-axes are stretched by a factor of $\sqrt{\frac{1+\beta^2}{1-\beta^2}}$. On the x- and x'-axes some tick marks have been added as a reminder. Thus, the length of the line segment from the origin to A has to be multiplied by that factor to obtain the same length in the S'-frame. It has thus a length of $cT'\sqrt{\frac{1+\beta^2}{1-\beta^2}}$. Now we can write with the Pythagorean theorem,

$$x_A^2 + (ct)_A^2 = \frac{\beta^2(cT)^2}{(1-\beta)^2} + \frac{(cT)^2}{(1-\beta)^2} = (cT')^2\frac{1+\beta^2}{1-\beta^2}$$

$$\frac{1+\beta^2}{(1-\beta)^2}T^2 = \frac{1+\beta^2}{1-\beta^2}T'^2$$

$$\frac{T}{T'} = \sqrt{\frac{(1-\beta)^2}{1-\beta^2}}$$

$$\frac{\nu'}{\nu} = \frac{T}{T'} = \sqrt{\frac{1-\beta}{1+\beta}}.$$

This is (14.3), Q.E.D.

The relativistic Doppler effect finds its most important *applications in astronomy*. It is sensible to assume that the contribution of the transverse Doppler effect to the total Doppler effect is negligible, i.e. that the measured Doppler effect can be attributed almost exclusively to radial velocities. Since in most disciplines of physics, such as optics, X-ray and radio-astronomy, wavelengths rather than frequencies are measured, the relativistic Doppler equation, (14.3), has to be *converted to wavelengths*. With $\nu = c/\lambda$, we obtain

$$\frac{\nu}{\nu_0} = \frac{\lambda_0}{\lambda} = \sqrt{\frac{1-\beta}{1+\beta}}$$

$$\left(\frac{\lambda_0}{\lambda}\right)^2 = \frac{1-\beta}{1+\beta}$$

$$\beta\left[\left(\frac{\lambda_0}{\lambda}\right)^2 + 1\right] = 1 - \left(\frac{\lambda_0}{\lambda}\right)^2$$

$$\beta = \frac{1 - \left(\dfrac{\lambda_0}{\lambda}\right)^2}{1 + \left(\dfrac{\lambda_0}{\lambda}\right)^2}. \tag{14.7}$$

From the shift of a characteristic spectral line by $\Delta\lambda = \lambda - \lambda_0$ (mostly $\Delta\lambda > 0$, i.e. a *red shift*), the relative velocity β of the cosmic source can immediately be determined. For those cases with a *blue shift* ($\Delta\lambda < 0$), the relative velocities are negative; these objects are in motion toward earth.

Even very small velocities, down to only several m/s, give rise to Doppler effects that can be measured today. This ability has become the foundation of a novel discipline of astronomy, the *discovery of extrasolar planets*, i.e. planets of distant stars. To explain this, we reduce our own solar system to a two-body system consisting of the sun ($M = 2.0 \times 10^{30}$ kg) and Jupiter ($m = 1.9 \times 10^{27}$ kg), orbiting the sun on a circle at a distance of $r = 0.78 \times 10^{12}$ m and at a velocity of $v_m = 13 \times 10^3$ m/s. So one Jovian year lasts $T = 12$ a. However, both Jupiter and the sun move around their common center of mass, located at 0.74×10^9 m from the center of the sun, i.e. close to the surface of the sun. The sun's velocity around that center of mass is $v_M = 12$ m/s.

An extraterrestrial observer whose direction of observation happens to be parallel to the plane of Jupiter's orbit cannot observe Jupiter with his instruments. However, he can notice that a star, called *sun*, is subject to a periodic motion with a period of $T = 12$ a, and an amplitude of $v_M = 12$ m/s which is attributable to the existence of an unknown companion. Since he is able to estimate the mass of sun with the known techniques of photometry (and spectroscopy), he can calculate the radius of the circular orbit of sun's companion to be

$$r = \left(\frac{GMT^2}{4\pi^2} \right)^{3/2},$$

with $G = 6.67 \times 10^{-11}$ m^3 kg^{-1}s^{-2} being the gravitational constant, and its mass to be

$$m = M \frac{v_M}{v_m} = M \frac{v_M T}{2\pi r}.$$

Chapter 15

The Twin Paradox and k-Calculus

Given are the twins A and B. While *A remains on earth, B is traveling* at a relativistic speed from earth to a nearby star, then back to earth. In the reference frame of A, B is subject to time dilation both on his journey toward the star and on his way back, so he ages more slowly and is younger than A when returning to earth. But in B's reference frame, A is subject to time dilation, so he should be younger than B at the time of their reunion. *After his journey, B seems to be both younger and older than A* – a typical paradox!

The paradox can easily be solved by noting that *the histories of A and B are not symmetric*: While A spends all his time in an inertial reference frame, namely the earth, B's journey includes *four stages of accelerated motion*: departure from earth (positive acceleration), arrival on the star (negative acceleration, also called deceleration), departure from the star (acceleration), and finally arrival on earth (deceleration). Consequently, A is living in a special reference frame which differs from B's reference frame by being an *inertial reference frame at all times*. The apparent symmetry between A and B has been broken! We will thus find B to be younger than A upon returning to earth.

One might argue that in B's reference frame A is subject to four accelerated phases, thus reestablishing a symmetry between A and B. This line of reasoning would be wrong! *Accelerations are not relative!* Whether a body is subject to acceleration or not can be determined *absolutely*, i.e. in a way valid for all reference frames. For an accelerated body is acted upon by inertial forces, while a non-accelerated body is not. This has the following consequences for the twins: During the two phases of positive acceleration, B is being pressed into the seat of his spaceship, and B has to use his safety belt during the two phases of negative acceleration. A experiences nothing of this kind. Accelerated motions are not relative like rectilinear uniform motions.

A numerical example will clarify the situation: B's destination is 8 light-years away, and the cruising speed amounts to $\beta = 0.8$. In order to avoid complications, we specify that the four accelerated phases of his journey are short. B's departure occurs in the year 2000. For A, B's outward voyage and return voyage require 10 years each, B returns in 2020. For B, the destination is moving at $\beta = -0.8$ toward him, so the distance from earth to the star is subject to length contraction and is reduced from 8 light-years to $8\sqrt{1 - (-0.8)^2} = 4.8$ light-years. So the outward voyage takes only $4.8\,\mathrm{ly}/0.8c = 6$ years. B marks 2006 as the year of arrival on his calendar. The return voyage is similar, thus B marks 2012 for the reunion on earth on his calendar. Obviously, B ages 8 years less than A.

How does A interpret this phenomenon? For him, the distance to the star is certainly not subject to length contraction, but B is subject to time dilation. So in 10 years on earth, B experiences only $10\sqrt{1 - 0.8^2} = 6$ years, which is why B reaches the star in 2006 by his calendar and returns to earth in 2012.

When length contraction and time dilation are under discussion, the impression may arise that an observer on earth looking at a spaceship passing earth at a velocity of β can actually *see* how the spaceship is being contracted by a factor of $\sqrt{1 - \beta^2}$, and can actually *see* how the lapse of time on a spaceship clock is being dilated by a factor of $1/\sqrt{1 - \beta^2}$. This impression is wrong because such considerations also require that the time needed for light to travel from the spaceship to the observer is included. After all, *seeing means receiving light from the object observed*. The question of what a fast moving, i.e. length-contracted, body really looks like will be dealt with in the next chapter. The point in this chapter is to calculate what the flow of time on a fast moving, i.e. time-dilated, calendar looks like, in other words how fast time seems to run.

According to A's calendar, B arrives on the star in the year 2010. Light from B's phosphorescent calendar-wristwatch takes another 8 years to reach earth and arrives there in 2018 of A's count. So in 2018, A sees B's calendar show 2006 and infers that all of B's movements – during his entire outward voyage – are being slowed by a factor of 3, not by a factor of $1/\sqrt{1 - 0.8^2} = 5/3$! This result is not new because when light from B's wristwatch is analyzed spectroscopically, the astronomer finds a frequency ratio of 1:3, or a lengthening of all periods by a factor of 3, according to the Doppler equation (14.3).

B also watches a calendar on earth. The terrestrial calendar display 2002 arrives on the star in 2010 of A's count and can be observed at the time of B's arrival, i.e. in 2006 of B's count. So B infers that all movements on earth – during his entire outward voyage – are also being slowed by a factor of 3. This has to be so, of course, as the Doppler effect is symmetric, i.e. it depends only on the relative velocity between light source and light receiver.

The departure for the return voyage is observed in 2018 (earth count) and the arrival in 2020 (earth count). A can observe B's calendar leap from 2006 to 2012. Thus, A can see B's movements being accelerated by a factor of 3. Not in the least does he observe a slowing down by a time dilation factor of 5/3! Of course, the Doppler equation (14.3) yields such a frequency ratio of 3:1 ($\beta = -0.8$), i.e. a shortening of all periods by a factor of 3.

Finally, B watches a calendar on earth on his way home. Upon departure from the star, it shows 2002, on his arrival on earth, it stands at 2020. His own calendar proceeds from 2006 to 2012, so he also observes all movements on earth being accelerated by a factor of 3. Of course, (14.3) is in accordance with this.

Minkowski diagrams of B's voyage are shown in Fig. 15.1. Twin A is at rest in the S-frame, twin B in the S'-frame; thus B's world line is the t'-axis. For better legibility, the vertical axes are denoted by t and t', rather than ct and ct'.

With Fig. 15.1, we can immediately understand *how the speed of a moving object can be measured by radar*. Upper diagram: A signal emitted by A in 2000 is immediately reflected by B and received by A, i.e. in 2000. A signal emitted in 2002 is received in 2018. Signals emitted in a two-year rhythm return in an eighteen-year rhythm, thus the frequency of a

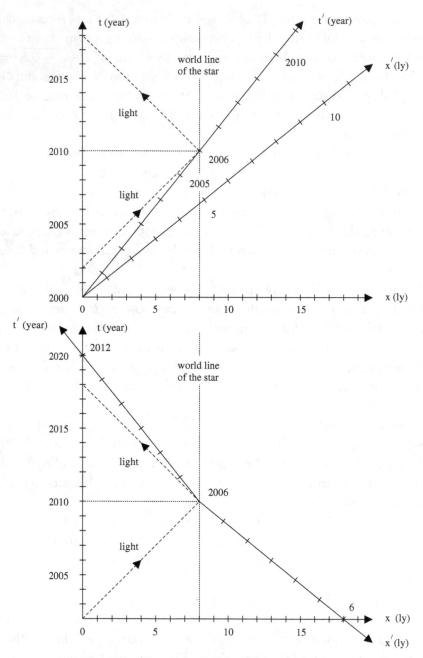

Fig. 15.1. Voyage of twin *B* in a Minkowski diagram:
voyage out (top) and voyage home (bottom).

signal received by A, after reflection at B, is reduced by a factor of 9. This is sensible as the frequency of a signal, when arriving at B, has already been reduced by a factor of 3, and is subject to yet another reduction by a factor of 3 through reflection at B. Lower diagram: A signal emitted in 2002 is received in 2018, a signal emitted in 2020 is returned at once. Signals emitted in an eighteen-year rhythm return in a two-year rhythm, thus the frequency is enhanced by a factor of 9. Of course, this result can be obtained by applying (14.3) twice, i.e.

$$\frac{\nu}{\nu_0} = \frac{1 - \beta}{1 + \beta}. \tag{15.1}$$

Since only non-relativistic speeds are measured on earth, a Taylor expansion of (15.1) to first order in β provides great accuracy, so

$$\frac{\nu}{\nu_0} \approx (1 - \beta)(1 - \beta) \approx 1 - 2\beta.$$

Actually, the relative frequency shift is measured,

$$\frac{\Delta \nu}{\nu_0} = \frac{\nu - \nu_0}{\nu_0} \approx -2\beta. \tag{15.2}$$

For $\nu = 30$ m/s, it is as small as -2×10^{-7}.

Considerations, similar to those about twins A and B, are the foundation of a new and independent method of deriving the Doppler equation and the Lorentz transformation. The idea goes back to *Hermann Bondi* and is called *k-calculus*, but can hardly be found in the literature in spite of its ingenious simplicity. Readers not interested in this detour may skip the rest of this chapter without losing the red thread.

Let us introduce the idea first with a numerical example similar to the one just used: Person A is at rest in the origin of the S-frame, let us say on the earth. Once a year, he sends a light flash to a star at rest relative to him, positioned 8 light-years to his right. The periodic light flashes arrive there, say at a receiver called C, in one-year intervals (Fig. 15.2).

Observer B, being at rest in the S'-frame, is moving from the common origin toward the star at a velocity of ν, the magnitude of which we pretend not to know at the moment. Every third year (B's time), he can see a light

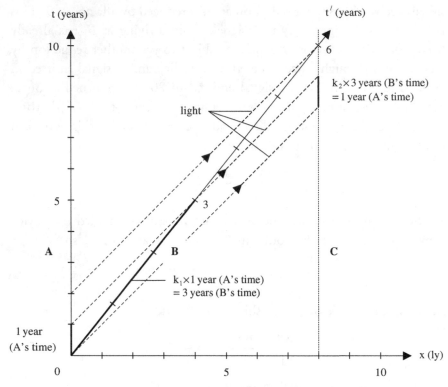

Fig. 15.2.   A moving observer, *B*, receiving and emitting
light flashes moving from left to right.

signal from *A* pass him. Let us imagine that, every time a light flash passes him, a second flash, maybe of different color, is emitted by *B* into the same direction. These two flashes propagate side by side and are simultaneously received by *C* in one-year intervals, as measured by *C*'s clock, running as fast as *A*'s clock.

Let us summarize: When the period of light emission at *A* (*A*'s time) is multiplied by $k_1 = 3$, we obtain the period of light reception at *B* (*B*'s time). When the period of light emission at *B* (*B*'s time) is multiplied by $k_2 = 1/3$, we get the period of light reception at *C* (*C*'s time = *A*'s time). Obviously, $k > 1$ pertains to $v > 0$ (increasing distance), and $k < 1$ to $v < 0$ (decreasing distance). This knowledge, however, is no prerequisite for what follows. How does *k* depend on the relative velocity *v*?

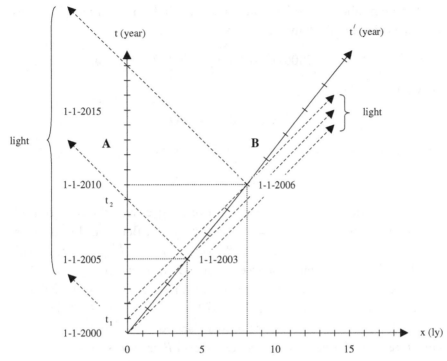

Fig. 15.3.   The moving observer B receives light flashes from
the left and re-emits light flashes to the left.

Let us continue this numerical example and change the situation slightly (Fig. 15.3): B leaves A on 1.1.2000 and reaches his cruising speed at once. A emits a light flash to his right on 1st January each year. Each time B receives such a light flash, he re-emits one to A. The light flash emitted on $t_1 = 1.1.2001$ (A's time) reaches B on 1.1.2003 (B's time), assuming $k = 3$. B immediately re-emits a light flash to A. Due to the symmetry between A and B – either may consider himself as resting and the other one as moving – $k = 3$ must also be true for B's light flash; so it reaches A on $t_2 = 1.1.2009$ (A's time). A's light flash to B, as well as the light flash at once re-emitted by B to A, are en route for a total of $t_2 - t_1 = 8$ years (A's time). The arrival of A's flash at B takes place $(t_2 - t_1)/2 = 4$ years after its emission, i.e. on 1.1.2005 (A's time). Thus, A and B are $c(t_2 - t_1)/2 = 4$ light-years apart (A's distance) on that day. Since A and B separated 5 years earlier (A's time), their relative velocity amounts to $4/5$ of c, or $\beta = 4/5$.

When the above considerations are written in formula language, things look like this: A emits light flashes on

$$t_1 = 1.1.2000 + \Delta t \quad (\Delta t = 0; 1; 2 \ldots \text{years}).$$

They arrive at B on

$$t' = 1.1.2000 + k\Delta t.$$

The light flashes emitted by B to his left reach A on

$$t_2 = 1.1.2000 + k^2 \Delta t.$$

For A, the period between the emission of a signal and the reception of an answer amounts to $t_2 - t_1 = (k^2 - 1)\Delta t$ years. Between the emission of a signal at A and its reception at B, $(t_2 - t_1)/2 = (k^2 - 1)\Delta t/2$ years have lapsed, so the distance between A and B amounts to $c(t_2 - t_1)/2 = (k^2 - 1)c\Delta t/2$ light-years. Of course, the instant of reception at B is halfway between t_1 and t_2, i.e. on $t = (t_1 + t_2)/2 = 1.1.2000 + (k^2 + 1)\Delta t/2$; on that day, A and B have already been separated for $(k^2 + 1)\Delta t/2$ years. Therefore, the relative velocity between A and B amounts to

$$v = \frac{(k^2 - 1)c\Delta t/2}{(k^2 + 1)\Delta t/2},$$

or

$$\beta = \frac{v}{c} = \frac{k^2 - 1}{k^2 + 1}. \tag{15.3}$$

Solving for k, we obtain

$$k = \sqrt{\frac{1 + \beta}{1 - \beta}}. \tag{15.4}$$

This is identical to (14.3) as $k = T/T_0 = v_0/v$. Since β is defined for the range from -1 to $+1$, the codomain of k ranges from 0 to ∞.

Now we can easily derive the Lorentz transform. For this purpose, we reintroduce ct and ct' as ordinates (Fig. 15.4). An observer at the origin of the S-frame is emitting a light flash at the moment $ct = ct_0 - x_0$. After a while, when this light flash passes the observer at the origin of the S'-frame,

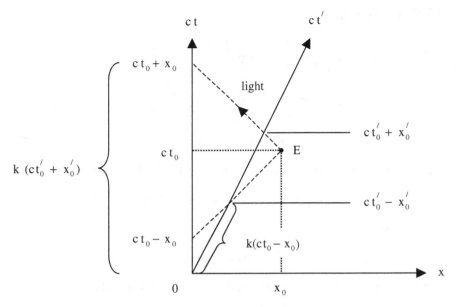

Fig. 15.4.   Derivation of the Lorentz transform with k-calculus.

he is also emitting a light flash. Both flashes propagate side by side to event E with the coordinates $(ct_0 \mid x_0)$ and are being reflected there. They return to the observer in the S'-frame and, somewhat later, to the observer in the S-frame at the moment $ct = ct_0 + x_0$.

The S'-observer assigns the coordinates $(ct_0' \mid x_0')$ to event E. As a result of the principle of special relativity, i.e. the *complete equivalence of both frames*, the mathematical formulation has to be *shape-invariant*. This means that – *except for the primes* – it has to be *identical in both frames*. Thus, in the S'-frame the emission is taking place at the moment $ct' = ct_0' - x_0'$, and the return of the flash at $ct' = ct_0' + x_0'$. With the ideas of k-calculus, the following two statements are true:

$$ct_0' - x_0' = k(ct_0 - x_0) \qquad (15.5)$$
$$ct_0 + x_0 = k(ct_0' + x_0'). \qquad (15.6)$$

We multiply (15.5) by $-k$, add (15.6), and obtain

$$2kct_0' = (k^2 + 1)ct_0 - (k^2 - 1)x_0.$$

(15.4) turns this into

$$2\sqrt{\frac{1+\beta}{1-\beta}}ct_0' = \frac{2}{1-\beta}ct_0 - \frac{2\beta}{1-\beta}x_0$$

$$ct_0' = \frac{1}{\sqrt{1-\beta^2}}ct_0 - \frac{\beta}{\sqrt{1-\beta^2}}x_0$$

$$ct_0' = \gamma(ct_0 - \beta x_0). \tag{15.7}$$

Correspondingly, we multiply (15.5) by k, add (15.6), and obtain

$$2kx_0' = (k^2 + 1)x_0 - (k^2 - 1)ct_0.$$

By analogy, this gives

$$x_0' = \gamma(x_0 - \beta ct_0). \tag{15.8}$$

Obviously, (15.7) and (15.8) are the Lorentz transforms, (6.5).

Chapter 16*

Images of Fast Moving Objects

Contrary to popular belief, the image of fast moving objects is determined not only by the phenomenon of length contraction, but also by the *transit-time that light requires to travel from the moving object to the observer*. The place where the object *is perceived* by the observer is *not* the place where the object *actually is* at the time of observation. This is called a *retardation effect*. The problem becomes even more intricate as points at opposite ends of the object have different transit-times. Therefore, the observer cannot even see the object as it appeared at a particular point in the past. On the contrary, each point of the image belongs to a different moment in the past. This gives rise to some rather peculiar distortions of the object's image, as we will find out in a moment.

Let us first assume that a light emitting object is *point-like*, e.g. the point in Fig. 6.1 designated as a five-pronged star. This point-like object is assumed to rest in the S'-frame at the locus (x', y', z'). Contrary to the considerations of Chapter 6, this point is *not an event*, i.e. not an instantaneous appearance, but it is *unlimited in time*. The observer of the object is assumed at rest in the origin of the S-frame. As usual, the axes of both frames are mutually parallel, S' is in motion relative to S at a velocity of $\mathbf{v} = (v, 0, 0)$, and the two origins coincide at $t = t' = 0$.

We want to use the S-frame for our calculations and first limit ourselves to an event, namely a light flash emitted by the object at the moment t_0 and at the locus (x, y, z). This light flash arrives in the origin at the moment t, with $t > t_0$. Since it has to travel a distance of $\sqrt{x^2 + y^2 + z^2}$, we find

$$t = t_0 + \frac{1}{c}\sqrt{x^2 + y^2 + z^2}. \tag{16.1}$$

For the x-coordinate of the event, the elementary relation (6.2) is true,

$$x = \frac{x'}{\gamma} + vt_0,$$

which was then used to derive the Lorentz transform. After rearrangement, it reads

$$ct_0 = \frac{1}{\beta}\left(x - \frac{x'}{\gamma}\right). \tag{16.2}$$

We also have $y = y'$ and $z = z'$.

Our task is to calculate the locus of emission in the observer's frame, (x, y, z), as a function of the locus of emission in the moving frame, (x', y', z'), for a certain arrival time t of the light flash. To do so, we insert (16.2) into (16.1), replace y and z by y' and z', respectively,

$$ct = \frac{1}{\beta}\left(x - \frac{x'}{\gamma}\right) + \sqrt{x^2 + y'^2 + z'^2},$$

rearrange terms and square,

$$c^2 t^2 + \frac{1}{\beta^2}x^2 + \frac{1}{\beta^2\gamma^2}x'^2 - \frac{2}{\beta}ctx + \frac{2}{\beta\gamma}ctx' - \frac{2}{\beta^2\gamma}xx' = x^2 + y'^2 + z'^2,$$

combine powers of x,

$$x^2\left(1 - \frac{1}{\beta^2}\right) + x\frac{2}{\beta}\left(\frac{x'}{\beta\gamma} + ct\right) + y'^2 + z'^2 - c^2t^2 - \frac{1}{\beta^2\gamma^2}x'^2 - \frac{2}{\beta\gamma}ctx' = 0,$$

and multiply by $-\beta^2\gamma^2$,

$$x^2 - 2x\gamma^2\left(\frac{x'}{\gamma} + \beta ct\right) - \beta^2\gamma^2(y'^2 + z'^2) + \gamma^2\left(\frac{x'}{\gamma} + \beta ct\right)^2 = 0.$$

The solutions of this quadratic equation in x are

$$x = \gamma^2\left(\frac{x'}{\gamma} + \beta ct\right) \pm \sqrt{(\gamma^4 - \gamma^2)\left(\frac{x'}{\gamma} + \beta ct\right)^2 + \beta^2\gamma^2(y'^2 + z'^2)},$$

or, with $\gamma^4 - \gamma^2 = \gamma^4 \left(1 - \frac{1}{\gamma^2}\right) = \gamma^4 \beta^2$,

$$x = \gamma^2 \left(\frac{x'}{\gamma} + \beta ct\right) \pm \beta \gamma \sqrt{\gamma^2 \left(\frac{x'}{\gamma} + \beta ct\right)^2 + y'^2 + z'^2}.$$

$\frac{x'}{\gamma} + \beta ct$ is the x-coordinate of the object when the flash arrives in the origin at t. Because the object continues to move on after emission, this must be greater than the x-coordinate of the object at the moment of emission, i.e. the left-hand side of the equation. Since $\gamma > 1$, the equation holds only for the negative sign, giving the final result for the coordinates of the locus of emission in the observer's frame,

$$x = \gamma(x' + \beta\gamma ct) - \beta\gamma\sqrt{(x' + \beta\gamma ct)^2 + y'^2 + z'^2}$$
$$y = y'$$
$$z = z'. \tag{16.3}$$

Now consider a light emitting object which is a lattice of 9 points grouped around the coordinate origin of the S'-frame, shown in Fig. 16.1. The distance between the points in the x'-direction and y'-direction is 1 light-second (ls) each, and for all points $z' = 0$ holds. At $t = t' = 0$, the center of the lattice is moving over the S-observer. The S'-frame moves at $\beta = 0.8$ relative to the S-frame.

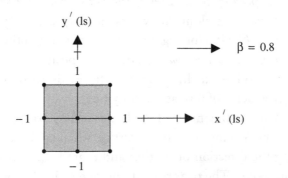

Fig. 16.1.   A lattice of 9 points in the S'-frame, the center of the lattice is in the coordinate origin.

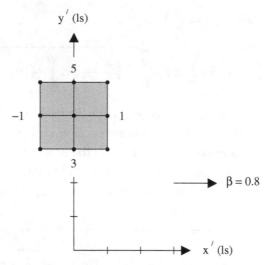

Fig. 16.2. A lattice of 9 points in the S'-frame, the center of the lattice is 4 ls away from the coordinate origin.

In a second calculation, the lattice is located in the S'-frame, as shown in Fig. 16.2. At $t = t' = 0$, its center is passing the observer at a distance of 4 light-seconds. The S'-frame again moves at $\beta = 0.8$ relative to the S-frame.

Figure 16.3 shows the image of the lattice of Fig. 16.1, i.e. what it would really look like, for an observer in the coordinate origin of the S-frame. On approaching, the lattice (white) *seems to be stretched* by a factor of 3 along the x-axis. The lattice *seems to move at 4 times*, instead of 0.8 times, *the speed of light*. After flying across the observer, it *appears shortened* by a factor of 3 along the x-axis, and *appears to depart at about 0.4 times the speed of light*. Line segments at right angles to the direction of motion *seem to be curved backwards*. The actual position of the lattice in the S-frame has also been included in gray. In the direction of motion, it is contracted by a factor of 0.6, as given by (4.3).

Figure 16.4 shows the image of the lattice of Fig. 16.2 for an observer in the origin of the S-frame. Again, the lattice, when approaching, is being stretched along the direction of motion and is gradually being shortened on crossing the y-axis. The distortion along one of its diagonals is slowly diminishing. The upper left corner becomes visible (see dotted lines) long

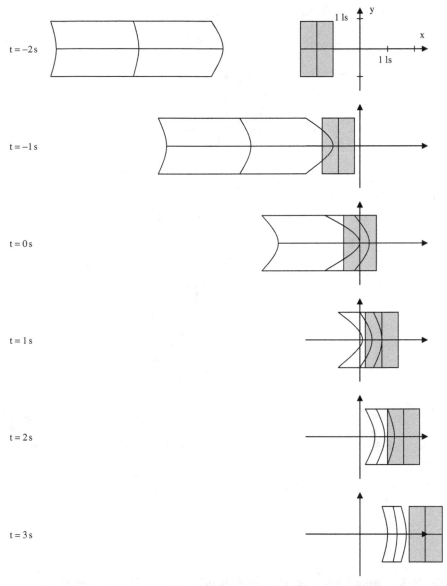

Fig. 16.3.   Lattice shown in Fig. 16.1, as seen in the S-frame, at different times. Gray: actual position of the contracted lattice; white: apparent shape and position, i.e. image, of the lattice.

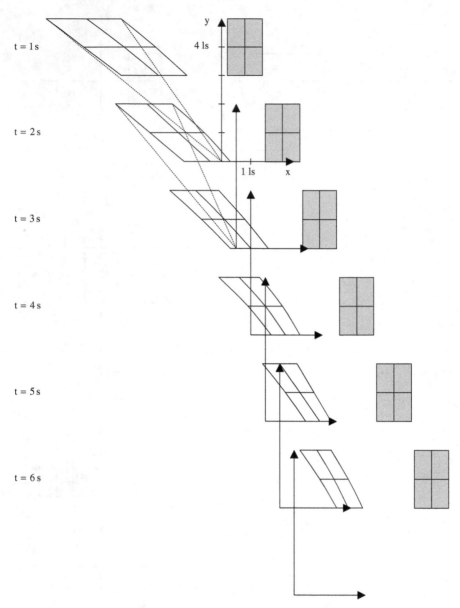

Fig. 16.4. Lattice shown in Fig. 16.2, as seen in the *S*-frame, at different times. Gray: actual position of the contracted lattice; white: apparent shape and position, i.e. image, of the lattice. For better legibility, the vertical coordinate axes have been displaced horizontally.

before the left edge of the lattice appears to have crossed the y-axis. Correspondingly, the upper right corner disappears prematurely (also shown with dotted lines).

We now want to explain these effects in detail. Let us first answer the question why distances along the direction of motion seem to be stretched or shortened (Fig. 16.5).

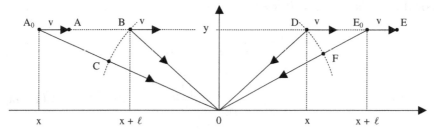

Fig. 16.5.   Apparent stretching (left) and shortening (right)
of a line segment along the direction of motion.

Left part of Fig. 16.5:

A_0 is the left end of the moving lattice at the time t_0. At that moment, light is being emitted from A_0. From A_0 to C it covers the distance

$$\sqrt{x^2 + y^2} - \sqrt{(x + \ell)^2 + y^2}. \quad (x < 0)$$

Thus, the light reaches C at the moment

$$t_0 + \frac{1}{c}\left(\sqrt{x^2 + y^2} - \sqrt{(x + \ell)^2 + y^2}\right).$$

At this moment, light is being emitted from the right end of the lattice which has arrived at B. It reaches the origin at the same instant as light from A_0. When light starts propagating from B, the left end has already moved on to A. The length of the line segment $A_0 A$ is

$$\frac{v}{c}\left(\sqrt{x^2 + y^2} - \sqrt{(x + \ell)^2 + y^2}\right).$$

Thus, in the S-frame, the length of the line segment AB is

$$\ell - \beta\left(\sqrt{x^2 + y^2} - \sqrt{(x + \ell)^2 + y^2}\right).$$

In the S'-frame, moving at $\mathbf{v} = (v, 0, 0)$, $(AB)'$ is dilated by a factor of γ. Thus, the ratio of the length *seen* in the S-frame, $\ell = A_0 B$, to the proper length *measured* in the S'-frame, $(AB)'$, is

$$\frac{\ell}{\gamma\left[\ell - \beta\left(\sqrt{x^2 + y^2} - \sqrt{(x + \ell)^2 + y^2}\right)\right]}.$$

For the special case $y = 0$, we find

$$\frac{A_0 B}{(AB)'} = \frac{\ell}{\gamma[\ell - \beta(|x| - |x + \ell|)]}$$

$$= \frac{1}{\gamma(1 - \beta)} = \sqrt{\frac{1 + \beta}{1 - \beta}}. \qquad (16.4a)$$

Right part of Fig. 16.5:

E_0 is the right end of the moving lattice at the time t_0. At that moment, light is being emitted from E_0. From E_0 to F it covers the distance

$$\sqrt{(x + \ell)^2 + y^2} - \sqrt{x^2 + y^2}. \quad (x > 0)$$

Thus, the light reaches F at the moment

$$t_0 + \frac{1}{c}\left(\sqrt{(x + \ell)^2 + y^2} - \sqrt{x^2 + y^2}\right).$$

At this moment, light is being emitted from the left end of the lattice which has arrived at D. It reaches the origin at the same instant as light from E_0. When light starts propagating from D, the right end has already moved on to E. The length of the line segment $E_0 E$ is

$$\frac{v}{c}\left(\sqrt{(x + \ell)^2 + y^2} - \sqrt{x^2 + y^2}\right).$$

Thus, in the S-frame, the length of the line segment DE is

$$\ell + \beta\left(\sqrt{(x + \ell)^2 + y^2} - \sqrt{x^2 + y^2}\right).$$

In the S'-frame, moving at $\mathbf{v} = (v, 0, 0)$, $(DE)'$ is dilated by a factor of γ. Thus, the ratio of the length *seen* in the S-frame, $\ell = DE_0$, to the proper length *measured* in the S'-frame, $(DE)'$, is

$$\frac{\ell}{\gamma[\ell + \beta(\sqrt{(x+\ell)^2 + y^2} - \sqrt{x^2 + y^2})]}.$$

For the special case $y = 0$, we find

$$\frac{DE_0}{(DE)'} = \frac{\ell}{\gamma[\ell + \beta(x + \ell - x)]}$$

$$= \frac{1}{\gamma(1 + \beta)} = \sqrt{\frac{1 - \beta}{1 + \beta}}. \qquad (16.4b)$$

(16.4a) and (16.4b) have exactly the structure of the Doppler equation (14.3) and is thus called the *length-Doppler effect*. Upon arrival and departure at a velocity of $\beta = 0.8$, the lattice seems to be stretched and shortened, respectively, by a factor of 3, as shown in Fig. 16.3.

In a similar manner, a *velocity-Doppler effect* can be calculated (Fig. 16.6). A lattice point A_0 with the coordinates $(x \mid y)$ is emitting light at the moment t_0. The light reaches the origin at

$$t_1 = t_0 + \frac{1}{c}\sqrt{x^2 + y^2}.$$

At time $t_0 + \frac{1}{v}dx$, the same lattice point A is emitting light at the coordinates $(x + dx \mid y)$ which reaches the origin at

$$t_2 = t_0 + \frac{1}{v}dx + \frac{1}{c}\sqrt{(x + dx)^2 + y^2}.$$

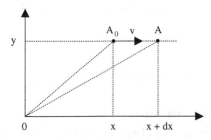

Fig. 16.6.   Calculating the velocity-Doppler effect.

Therefore, the elapsed time between the two flashes arriving at the origin is

$$dt = t_2 - t_1 = \frac{1}{v}dx + \frac{1}{c}\sqrt{(x + dx)^2 + y^2} - \frac{1}{c}\sqrt{x^2 + y^2}.$$

The first square root is expanded to give

$$\sqrt{(x + dx)^2 + y^2} = \sqrt{x^2\left(1 + \frac{dx}{x}\right)^2 + y^2} = \sqrt{x^2 + 2xdx + y^2}.$$

This Taylor expansion, even though it is only to first order, is not an approximation, but an exact expression because xdx, containing a differential quantity, is already infinitesimally small compared with $x^2 + y^2$. A second Taylor expansion, also being exact, gives

$$\sqrt{x^2 + y^2 + 2xdx} = \sqrt{x^2 + y^2}\sqrt{1 + \frac{2xdx}{x^2 + y^2}}$$

$$= \sqrt{x^2 + y^2}\left(1 + \frac{xdx}{x^2 + y^2}\right).$$

This yields

$$dt = \frac{1}{v}dx + \frac{1}{c}\frac{xdx}{\sqrt{x^2 + y^2}}.$$

The apparent velocity v^* of that point is

$$v^* = \frac{dx}{dt} = \frac{1}{\dfrac{1}{v} + \dfrac{1}{c}\dfrac{x}{\sqrt{x^2 + y^2}}} = \frac{v}{1 + \dfrac{\beta x}{\sqrt{x^2 + y^2}}}. \tag{16.5}$$

The special case $y = 0$ gives

$$v^* = \frac{v}{1 - \beta} \quad (x < 0) \quad \text{and} \quad v^* = \frac{v}{1 + \beta} \quad (x > 0). \tag{16.6}$$

When the lattice is approaching, $v^* = 5v = 4c$ holds, when it departs, $v^* = 5v/9 = 4c/9$ holds. So (16.6) confirms the values we found directly from Figure 16.3.

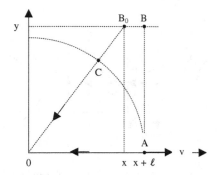

Fig. 16.7. Calculating the apparent backward curving of line segments at right angles to the direction of motion.

Next we examine why *line segments* at right angles to the direction of motion *appear to be curved* (Fig. 16.7). A and B are two points of the lattice with identical x-coordinates. At the moment t_0, light is being emitted from B_0. In the time it takes B_0 to move the distance ℓ to B, the light covers a distance of

$$\sqrt{x^2 + y^2} - |x + \ell|$$

to point C, reaching C at

$$t_0 + \frac{1}{c}\left(\sqrt{x^2 + y^2} - |x + \ell|\right).$$

At this moment, light is being emitted from A which will reach the origin at the same instant as light from B_0. The length of the line segment B_0B is

$$\frac{v}{c}\left(\sqrt{x^2 + y^2} - |x + \ell|\right).$$

On the other hand, B_0B is also equal to ℓ, i.e. equal to the distance that point B *seems* to lag behind point A. This gives

$$\ell = \beta\left(\sqrt{x^2 + y^2} - |x + \ell|\right),$$

and allows a determination of ℓ.

Finally, we would like to understand why *light can be seen from the back* of a fast moving object (Fig. 16.8). Left: a square object is at rest. A light emitting point on its left is visible only in the half-space to its left.

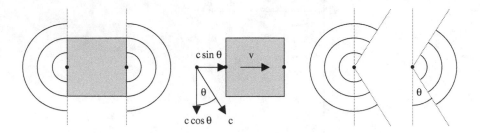

Fig. 16.8. Why the back of an object can be seen.

A corresponding statement is true for a light emitting point on its right (i.e. its front). Center: the object is moving at a velocity of v toward the observer. Light emitted by the point on the left (back of the object) has a velocity component to the right of $c \sin \theta \leq v$; thus $\theta \leq \arcsin \beta$. Right: light emitted on the left is visible in two additional sectors of width θ each. Correspondingly, light emitted by the point on the right can only be seen in a sector of width $\pi - 2\theta$. *The back appears early, the front disappears early.*

Chapter 17*

Rotation and Lorentz Transformation

A conspicuous formal similarity exists between the rotation of a Cartesian coordinate frame and a Lorentz transformation, as visualized by a Minkowski diagram.

Figure 17.1 shows the *rotation of a Cartesian coordinate frame* by an angle of δ. The coordinates $(x \mid y)$ of point P are to be transformed into the coordinates $(x' \mid y')$ of the rotated coordinate frame. One can immediately read from the figure

$$x' = r\cos(\varphi - \delta) = r\cos\varphi\cos\delta + r\sin\varphi\sin\delta = x\cos\delta + y\sin\delta$$
$$y' = r\sin(\varphi - \delta) = r\sin\varphi\cos\delta - r\cos\varphi\sin\delta = y\cos\delta - x\sin\delta.$$

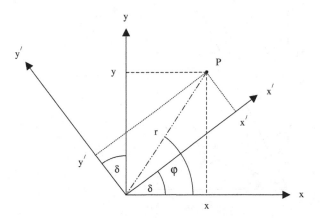

Fig. 17.1. Rotation of a Cartesian coordinate frame.

Written with a matrix,

$$\begin{pmatrix} x' \\ y' \end{pmatrix} = \begin{pmatrix} \cos\delta & \sin\delta \\ -\sin\delta & \cos\delta \end{pmatrix} \begin{pmatrix} x \\ y \end{pmatrix}. \tag{17.1}$$

Figure 17.2 shows a *Minkowski diagram*. The transformation of the coordinates $(ct \mid x)$ of an event E into the coordinates $(ct' \mid x')$ can also be written with a matrix,

$$\begin{pmatrix} ct' \\ x' \end{pmatrix} = \begin{pmatrix} \cosh\varphi & -\sinh\varphi \\ -\sinh\varphi & \cosh\varphi \end{pmatrix} \begin{pmatrix} ct \\ x \end{pmatrix}. \tag{17.2}$$

φ, the so-called *rapidity*, is given by

$$\varphi = \ln\sqrt{\frac{1+\beta}{1-\beta}}, \quad \text{with} \quad \beta = \frac{v}{c}. \tag{17.3}$$

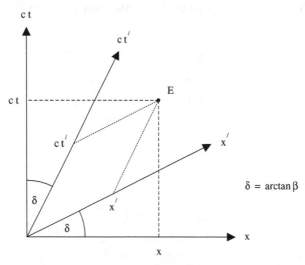

Fig. 17.2.   A Minkowski diagram.

To prove (17.2), we replace the hyperbolic functions by exponential functions,

$$\sinh\left(\ln\sqrt{\frac{1+\beta}{1-\beta}}\right) = \frac{1}{2}e^{\ln\sqrt{}} - \frac{1}{2}e^{-\ln\sqrt{}}$$

$$= \frac{\sqrt{1+\beta}}{2\sqrt{1-\beta}} - \frac{\sqrt{1-\beta}}{2\sqrt{1+\beta}} = \frac{(1+\beta)-(1-\beta)}{2\sqrt{1-\beta^2}} = \gamma\beta$$

$$\cosh\left(\ln\sqrt{\frac{1+\beta}{1-\beta}}\right) = \frac{1}{2}e^{\ln\sqrt{}} + \frac{1}{2}e^{-\ln\sqrt{}}$$

$$= \frac{\sqrt{1+\beta}}{2\sqrt{1-\beta}} + \frac{\sqrt{1-\beta}}{2\sqrt{1+\beta}} = \frac{(1+\beta)+(1-\beta)}{2\sqrt{1-\beta^2}} = \gamma,$$

and write

$$\begin{pmatrix} ct' \\ x' \end{pmatrix} = \begin{pmatrix} \gamma & -\gamma\beta \\ -\gamma\beta & \gamma \end{pmatrix} \begin{pmatrix} ct \\ x \end{pmatrix}, \tag{17.4}$$

or

$$ct' = \gamma ct - \gamma\beta x$$
$$x' = -\gamma\beta ct + \gamma x.$$

This is the Lorentz transformation, (6.5), Q.E.D.

We now want to apply *two Lorentz transformations in succession*. First, however, we limit ourselves to the case that *both transformations are in the x-direction*. We write

$$1^{\text{st}}\text{ LT}: \varphi_1 = \ln\sqrt{\frac{1+\beta_1}{1-\beta_1}}, \quad \text{and} \quad 2^{\text{nd}}\text{ LT}: \varphi_2 = \ln\sqrt{\frac{1+\beta_2}{1-\beta_2}}.$$

The first and the second Lorentz transformation, respectively, are given by

$$\begin{pmatrix} ct' \\ x' \end{pmatrix} = \begin{pmatrix} \cosh \varphi_1 & -\sinh \varphi_1 \\ -\sinh \varphi_1 & \cosh \varphi_1 \end{pmatrix} \begin{pmatrix} ct \\ x \end{pmatrix}$$

$$\begin{pmatrix} ct'' \\ x'' \end{pmatrix} = \begin{pmatrix} \cosh \varphi_2 & -\sinh \varphi_2 \\ -\sinh \varphi_2 & \cosh \varphi_2 \end{pmatrix} \begin{pmatrix} ct' \\ x' \end{pmatrix},$$

or combined,

$$\begin{pmatrix} ct'' \\ x'' \end{pmatrix} = \begin{pmatrix} \cosh \varphi_2 & -\sinh \varphi_2 \\ -\sinh \varphi_2 & \cosh \varphi_2 \end{pmatrix} \begin{pmatrix} \cosh \varphi_1 & -\sinh \varphi_1 \\ -\sinh \varphi_1 & \cosh \varphi_1 \end{pmatrix} \begin{pmatrix} ct \\ x \end{pmatrix}.$$

With the addition theorems for hyperbolic functions,

$$\sinh(\varphi_1 + \varphi_2) = \sinh \varphi_1 \cosh \varphi_2 + \cosh \varphi_1 \sinh \varphi_2$$
$$\cosh(\varphi_1 + \varphi_2) = \cosh \varphi_1 \cosh \varphi_2 + \sinh \varphi_1 \sinh \varphi_2,$$

we obtain

$$\begin{pmatrix} ct'' \\ x'' \end{pmatrix} = \begin{pmatrix} \cosh(\varphi_1 + \varphi_2) & -\sinh(\varphi_1 + \varphi_2) \\ -\sinh(\varphi_1 + \varphi_2) & \cosh(\varphi_1 + \varphi_2) \end{pmatrix} \begin{pmatrix} ct \\ x \end{pmatrix}. \tag{17.5}$$

Obviously, two Lorentz transformations in series, (17.5), exhibit exactly the same structure as one transformation alone, (17.2). One must not attempt to calculate $\beta_{1+2} = \beta_1 + \beta_2$, as we already know from the addition theorem of velocities, but rather $\varphi_{1+2} = \varphi_1 + \varphi_2$. What is left to be shown is:

$$\ln \sqrt{\frac{1 + \beta_{1+2}}{1 - \beta_{1+2}}} \overset{!}{=} \ln \sqrt{\frac{1 + \beta_1}{1 - \beta_1}} + \ln \sqrt{\frac{1 + \beta_2}{1 - \beta_2}}.$$

Proof: Exponentiation and squaring gives

$$\frac{1 + \beta_{1+2}}{1 - \beta_{1+2}} = \frac{1 + \beta_1}{1 - \beta_1} \frac{1 + \beta_2}{1 - \beta_2},$$

or solved for β_{1+2},

$$\beta_{1+2} \left[1 + \frac{(1 + \beta_1)(1 + \beta_2)}{(1 - \beta_1)(1 - \beta_2)} \right] = \frac{(1 + \beta_1)(1 + \beta_2)}{(1 - \beta_1)(1 - \beta_2)} - 1$$

$$\beta_{1+2} = \frac{(1 + \beta_1)(1 + \beta_2) - (1 - \beta_1)(1 - \beta_2)}{(1 - \beta_1)(1 - \beta_2) + (1 + \beta_1)(1 + \beta_2)} = \frac{\beta_1 + \beta_2}{1 + \beta_1 \beta_2}.$$

This is the addition theorem of velocities, (9.1), Q.E.D.

It is interesting to find out *whether two or more Lorentz transformations in succession have to be applied in a given order.*

$$\mathbf{L}_1^x = \begin{pmatrix} \gamma_1 & -\gamma_1\beta_1 \\ -\gamma_1\beta_1 & \gamma_1 \end{pmatrix}$$

and

$$\mathbf{L}_2^x = \begin{pmatrix} \gamma_2 & -\gamma_2\beta_2 \\ -\gamma_2\beta_2 & \gamma_2 \end{pmatrix}$$

denote two Lorentz transformations, pertaining to the relative velocities $\mathbf{v}_1 = (v_1, 0, 0)$ and $\mathbf{v}_2 = (v_2, 0, 0)$, respectively. If we first transform from an S-frame to an S'-frame, moving at $\mathbf{v}_1$ relative to the S-frame, and thereafter to an S''-frame, moving at $\mathbf{v}_2$ relative to the S'-frame, the mathematical operation is

$$\begin{pmatrix} ct'' \\ x'' \end{pmatrix} = \begin{pmatrix} \gamma_2 & -\gamma_2\beta_2 \\ -\gamma_2\beta_2 & \gamma_2 \end{pmatrix} \begin{pmatrix} \gamma_1 & -\gamma_1\beta_1 \\ -\gamma_1\beta_1 & \gamma_1 \end{pmatrix} \begin{pmatrix} ct \\ x \end{pmatrix}$$

$$= \begin{pmatrix} \gamma_1\gamma_2(1+\beta_1\beta_2) & -\gamma_1\gamma_2(\beta_1+\beta_2) \\ -\gamma_1\gamma_2(\beta_1+\beta_2) & \gamma_1\gamma_2(1+\beta_1\beta_2) \end{pmatrix} \begin{pmatrix} ct \\ x \end{pmatrix}.$$

If transformation $\mathbf{L}_2^x$ is applied before transformation $\mathbf{L}_1^x$, we have

$$\begin{pmatrix} ct'' \\ x'' \end{pmatrix} = \begin{pmatrix} \gamma_1 & -\gamma_1\beta_1 \\ -\gamma_1\beta_1 & \gamma_1 \end{pmatrix} \begin{pmatrix} \gamma_2 & -\gamma_2\beta_2 \\ -\gamma_2\beta_2 & \gamma_2 \end{pmatrix} \begin{pmatrix} ct \\ x \end{pmatrix}$$

$$= \begin{pmatrix} \gamma_1\gamma_2(1+\beta_1\beta_2) & -\gamma_1\gamma_2(\beta_1+\beta_2) \\ -\gamma_1\gamma_2(\beta_1+\beta_2) & \gamma_1\gamma_2(1+\beta_1\beta_2) \end{pmatrix} \begin{pmatrix} ct \\ x \end{pmatrix}.$$

Both operations give identical results, so

$$\mathbf{L}_1^x\mathbf{L}_2^x - \mathbf{L}_2^x\mathbf{L}_1^x = 0. \tag{17.6}$$

There is no order for applying *two Lorentz transformations into the same direction*; they are said to *commute*. Of course, the commutability of Lorentz transformations is immediately obvious from (17.5) as interchanging φ_1 and φ_2 does not alter (17.5).

What about *Lorentz transformations into two different directions*, e.g. L_1^x into the x-direction, $\mathbf{v}_1 = (v_1, 0, 0)$, and L_2^y into the y-direction, $\mathbf{v}_2 = (0, v_2, 0)$? We calculate

$$L_1^x L_2^y = \begin{pmatrix} \gamma_1 & -\gamma_1\beta_1 & 0 \\ -\gamma_1\beta_1 & \gamma_1 & 0 \\ 0 & 0 & 1 \end{pmatrix} \begin{pmatrix} \gamma_2 & 0 & -\gamma_2\beta_2 \\ 0 & 1 & 0 \\ -\gamma_2\beta_2 & 0 & \gamma_2 \end{pmatrix}$$

$$= \begin{pmatrix} \gamma_1\gamma_2 & -\gamma_1\beta_1 & -\gamma_1\gamma_2\beta_2 \\ -\gamma_1\gamma_2\beta_1 & \gamma_1 & \gamma_1\gamma_2\beta_1\beta_2 \\ -\gamma_2\beta_2 & 0 & \gamma_2 \end{pmatrix},$$

and

$$L_2^y L_1^x = \begin{pmatrix} \gamma_2 & 0 & -\gamma_2\beta_2 \\ 0 & 1 & 0 \\ -\gamma_2\beta_2 & 0 & \gamma_2 \end{pmatrix} \begin{pmatrix} \gamma_1 & -\gamma_1\beta_1 & 0 \\ -\gamma_1\beta_1 & \gamma_1 & 0 \\ 0 & 0 & 1 \end{pmatrix}$$

$$= \begin{pmatrix} \gamma_1\gamma_2 & -\gamma_1\gamma_2\beta_1 & -\gamma_2\beta_2 \\ -\gamma_1\beta_1 & \gamma_1 & 0 \\ -\gamma_1\gamma_2\beta_2 & \gamma_1\gamma_2\beta_1\beta_2 & \gamma_2 \end{pmatrix},$$

and find that the transformations *do not commute*, or

$$L_1^x L_2^y - L_2^y L_1^x \neq 0. \tag{17.7}$$

We see that *great care is needed with transformations into two different directions*, say x and y. First applying a Lorentz transformation into the x-direction, then another one into the y-direction cannot be the solution because, for reasons of symmetry, two transformations in reversed order should give the same result. But the results differ! When a motion in two directions is under discussion, the more complicated formalism of the following chapter *has* to be applied.

Chapter 18*

Lorentz Transformation in Two and Three Spatial Directions

Up to here, the situation has always been this: The S-frame and the S'-frame are coordinate frames with mutually parallel axes the origins of which coincide at $t = t' = 0$. The S'-frame moves at the constant velocity $\mathbf{v} = (v, 0, 0)$ relative to the S-frame, i.e. along its x-axis. The Lorentz transformation, or the inverse Lorentz transformation, is used to convert the four coordinates of an event from one frame into the other one.

But what if *the geometry is more complicated*, such that the S'-frame has to have the velocity $\mathbf{v} = (v_x, v_y, 0)$ relative to the S-frame? One may think of the following situation: The coordinates of an event E are to be described by three observers, A, B, and C. A is at rest in the S-frame, B is in motion relative to A at $\mathbf{v_B} = (v, 0, 0)$, and C is in motion relative to A at $\mathbf{v_C} = (v_x, v_y, 0)$. B and C are also at rest in their respective coordinate frames whose axes are mutually parallel to those of the S-frame and coincide at the moment $t_A = t_B = t_C = 0$ (Fig. 18.1).

The coordinates of the S-frame can always be spanned in such a way that the second observer, B, moves along the x-axis. However, the motion of the third observer, C, has at least two components, e.g. x and y. To calculate a *Lorentz transformation in two spatial directions*, one first has to rotate the S-frame by an angle δ into an S^*-frame for the velocity vector $\mathbf{v_C}$ to become parallel to the x^*-axis (Fig. 18.2). Mathematically speaking, the four frame (ct, x, y, z) is multiplied by the rotational matrix, (17.1), yielding a new four frame, (ct^*, x^*, y^*, z^*),

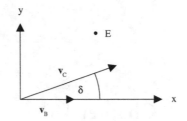

Fig. 18.1. Motion of *B* and *C* relative to *A*.

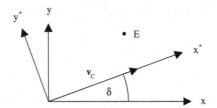

Fig. 18.2. Rotation of the *S*-frame into an *S**-frame.

$$\begin{pmatrix} ct^* \\ x^* \\ y^* \\ z^* \end{pmatrix} = \begin{pmatrix} 1 & 0 & 0 & 0 \\ 0 & \cos\delta & \sin\delta & 0 \\ 0 & -\sin\delta & \cos\delta & 0 \\ 0 & 0 & 0 & 1 \end{pmatrix} \begin{pmatrix} ct \\ x \\ y \\ z \end{pmatrix} = \begin{pmatrix} ct \\ x\cos\delta + y\sin\delta \\ -x\sin\delta + y\cos\delta \\ z \end{pmatrix}.$$

Subsequently, the Lorentz transformation (17.4) is applied to convert the *S**-frame into an *S***-frame with

$$\gamma = \frac{1}{\sqrt{1-\beta^2}}; \quad \beta = \frac{\sqrt{v_x^2 + v_y^2}}{c} = \frac{v_c}{c}. \tag{18.1}$$

The result is

$$\begin{pmatrix} ct^{**} \\ x^{**} \\ y^{**} \\ z^{**} \end{pmatrix} = \begin{pmatrix} \gamma & -\gamma\beta & 0 & 0 \\ -\gamma\beta & \gamma & 0 & 0 \\ 0 & 0 & 1 & 0 \\ 0 & 0 & 0 & 1 \end{pmatrix} \begin{pmatrix} ct^* \\ x^* \\ y^* \\ z^* \end{pmatrix}$$

$$= \begin{pmatrix} \gamma & -\gamma\beta & 0 & 0 \\ -\gamma\beta & \gamma & 0 & 0 \\ 0 & 0 & 1 & 0 \\ 0 & 0 & 0 & 1 \end{pmatrix} \begin{pmatrix} ct \\ x\cos\delta + y\sin\delta \\ -x\sin\delta + y\cos\delta \\ z \end{pmatrix}$$

$$= \begin{pmatrix} \gamma ct - \gamma\beta x\cos\delta - \gamma\beta y\sin\delta \\ -\gamma\beta ct + \gamma x\cos\delta + \gamma y\sin\delta \\ -x\sin\delta + y\cos\delta \\ z \end{pmatrix} .$$

Finally, the S^{**}-frame has to be rotated back by an angle $-\delta$ into the S'-frame,

$$\begin{pmatrix} ct' \\ x' \\ y' \\ z' \end{pmatrix} = \begin{pmatrix} 1 & 0 & 0 & 0 \\ 0 & \cos(-\delta) & \sin(-\delta) & 0 \\ 0 & -\sin(-\delta) & \cos(-\delta) & 0 \\ 0 & 0 & 0 & 1 \end{pmatrix} \begin{pmatrix} ct^{**} \\ x^{**} \\ y^{**} \\ z^{**} \end{pmatrix}$$

$$= \begin{pmatrix} ct^{**} \\ x^{**}\cos\delta - y^{**}\sin\delta \\ x^{**}\sin\delta + y^{**}\cos\delta \\ z^{**} \end{pmatrix} ,$$

making use of the trigonometric relations $\cos(-\delta) = \cos\delta$ and $\sin(-\delta) = -\sin\delta$. The result is

$$\begin{pmatrix} ct' \\ x' \\ y' \\ z' \end{pmatrix} = \begin{pmatrix} \gamma ct - \gamma\beta x\cos\delta - \gamma\beta y\sin\delta \\ -\gamma\beta ct\cos\delta + \gamma x\cos^2\delta + \gamma y\sin\delta\cos\delta + x\sin^2\delta - y\sin\delta\cos\delta \\ -\gamma\beta ct\sin\delta + \gamma x\sin\delta\cos\delta + \gamma y\sin^2\delta - x\sin\delta\cos\delta + y\cos^2\delta \\ z \end{pmatrix} .$$

The right-hand side can be written as a product of a matrix and a four frame which gives us the desired Lorentz-Transformation,

$$\begin{pmatrix} ct' \\ x' \\ y' \\ z' \end{pmatrix} = \begin{pmatrix} \gamma & -\gamma\beta\cos\delta & -\gamma\beta\sin\delta & 0 \\ -\gamma\beta\cos\delta & \gamma\cos^2\delta + \sin^2\delta & (\gamma-1)\sin\delta\cos\delta & 0 \\ -\gamma\beta\sin\delta & (\gamma-1)\sin\delta\cos\delta & \gamma\sin^2\delta + \cos^2\delta & 0 \\ 0 & 0 & 0 & 1 \end{pmatrix} \begin{pmatrix} ct \\ x \\ y \\ z \end{pmatrix} .$$

$$(18.2)$$

This equation is the analogue of (17.4), with the exception that the velocity vector that transforms the S-frame into the S'-frame has been generalized from one (x) to two spatial directions (x and y).

The inverse Lorentz transformation is obtained in the usual manner by interchanging primed and unprimed variables and replacing β by $-\beta$,

$$
\begin{pmatrix} ct \\ x \\ y \\ z \end{pmatrix} = \begin{pmatrix} \gamma & \gamma\beta\cos\delta & \gamma\beta\sin\delta & 0 \\ \gamma\beta\cos\delta & \gamma\cos^2\delta + \sin^2\delta & (\gamma-1)\sin\delta\cos\delta & 0 \\ \gamma\beta\sin\delta & (\gamma-1)\sin\delta\cos\delta & \gamma\sin^2\delta + \cos^2\delta & 0 \\ 0 & 0 & 0 & 1 \end{pmatrix} \begin{pmatrix} ct' \\ x' \\ y' \\ z' \end{pmatrix}.
$$

$$(18.3)$$

The foregoing calculation was rather intricate and certainly non-trivial for beginners. To allay any doubts about the result, we would now like to apply some *plausibility checks* by reducing the general solution to some special cases with answers already known. If all tests are passed, we have an *argument for, but not a proof of,* the correctness of the solution. If one single test failed, however, the solution would be wrong.

The first plausibility check is the transition to the case of low velocities which turns the Lorentz transformation into a Galilei transformation. To do so, we expand the γ-factor,

$$
\gamma = (1-\beta^2)^{-1/2} = 1 + \frac{1}{2}\beta^2 + \cdots .
$$

Up to first order in β, we find $\gamma \approx 1$, and $\gamma\beta \approx \beta$. This turns (18.2) into

$$
\begin{pmatrix} ct' \\ x' \\ y' \\ z' \end{pmatrix} = \begin{pmatrix} 1 & -\beta\cos\delta & -\beta\sin\delta & 0 \\ -\beta\cos\delta & 1 & 0 & 0 \\ -\beta\sin\delta & 0 & 1 & 0 \\ 0 & 0 & 0 & 1 \end{pmatrix} \begin{pmatrix} ct \\ x \\ y \\ z \end{pmatrix}.
$$

Thus,

$$
ct' = ct - x\beta\cos\delta - y\beta\sin\delta
$$

$$
t' = t - \frac{v_x}{c^2}x - \frac{v_y}{c^2}y = t - \left(\frac{v_x}{c}\right)^2 \frac{x}{v_x} - \left(\frac{v_y}{c}\right)^2 \frac{y}{v_y}.
$$

The second and third summands contain only second order terms of v/c. To first order we obtain, as expected, $t' = t$. Also,

$$x' = -ct\beta \cos \delta + x = x - v_x t$$
$$y' = -ct\beta \sin \delta + y = y - v_y t.$$

The Galilei transformation is thus obtained for x' and y' as well.

In a second plausibility check of (18.2), we set the angle δ between the velocity vector and the x-axis equal to zero,

$$
\begin{pmatrix} ct' \\ x' \\ y' \\ z' \end{pmatrix} = \begin{pmatrix} \gamma & -\gamma\beta & 0 & 0 \\ -\gamma\beta & \gamma & 0 & 0 \\ 0 & 0 & 1 & 0 \\ 0 & 0 & 0 & 1 \end{pmatrix} \begin{pmatrix} ct \\ x \\ y \\ z \end{pmatrix}.
$$

This equation is (17.4), i.e. the Lorentz transformation in one spatial direction.

As a third check, let us transform the S-frame with $\mathbf{v} = (v_x, v_y, 0)$ into the S'-frame. If, subsequently, the S'-frame is subjected to a transformation with $-\mathbf{v} = (-v_x, -v_y, 0)$, we should return to the S-frame. To do this check, we insert the Lorentz transformation (18.2), called $\mathbf{L}$, into the inverse Lorentz transformation (18.3), called $\mathbf{L}^{-1}$, and obtain

$$
\begin{pmatrix} ct \\ x \\ y \\ z \end{pmatrix} = \mathbf{L}^{-1}\mathbf{L} \begin{pmatrix} ct \\ x \\ y \\ z \end{pmatrix}.
$$

$\mathbf{L}^{-1}\mathbf{L} = 1$, i.e. the *unit matrix*,

$$
1 = \begin{pmatrix} 1 & 0 & 0 & 0 \\ 0 & 1 & 0 & 0 \\ 0 & 0 & 1 & 0 \\ 0 & 0 & 0 & 1 \end{pmatrix},
$$

should be the result of this operation. For a change from our usual practice of solving all problems in detail, we would like to ask the reader to multiply these two four-by-four matrices and show that their product is, indeed, the unit matrix.

The final proof, however, for the Lorentz transformation in two spatial directions to be correct is found if we can show that the scalar product of the time-position vector with itself is a Lorentz invariant,

$$(ct')^2 - x'^2 - y'^2 - z'^2 = (ct)^2 - x^2 - y^2 - z^2.$$

What scalar products of four-vectors and Lorentz invariance are all about will be the subject matter in chapters to follow. The reader is asked, however, to work out this lengthy though elementary proof as an exercise.

Let us now turn to *Lorentz transformations in three spatial directions*. One might expect an even more horrendous mathematical formalism. This is, fortunately, not the case. We will rather find that the Lorentz transformation in two spatial directions could have been derived much more easily.

In the most general case the situation is the following: The time-space coordinates of an event E are to be described by observer A at rest at the origin of the S-frame, and by observer B at rest at the origin of the S'-frame, both frames having mutually parallel axes. The origins coincide at $t = t' = 0$. B moves relative to A at a velocity of $\mathbf{v} = (v_x, v_y, v_z)$. The position vector of the event E in the S-frame, $\mathbf{r} = (x, y, z)$, is decomposed into a vector parallel to $\mathbf{v}$, called $\mathbf{r}_{\parallel}$, and a vector perpendicular to it, called $\mathbf{r}_{\perp}$ (Fig. 18.3).

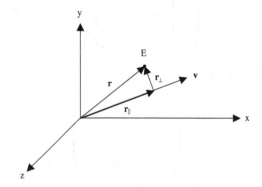

Fig. 18.3.   Decomposition of the position vector $\mathbf{r}$ into components parallel and perpendicular to $\mathbf{v}$.

Then the following Lorentz transformation is valid for the S'-frame with $\boldsymbol{\beta} = \mathbf{v}/c$,

$$ct' = \gamma(ct - \beta r_{\parallel}) = \gamma(ct - \boldsymbol{\beta} \cdot \mathbf{r})$$
$$\mathbf{r}'_{\parallel} = \gamma(\mathbf{r}_{\parallel} - \boldsymbol{\beta} ct)$$
$$\mathbf{r}'_{\perp} = \mathbf{r}_{\perp}.$$

The projection of the position vector $\mathbf{r}$ onto the velocity vector $\mathbf{v}$ is the scalar product of $\mathbf{r}$ with the unit vector $\hat{\mathbf{v}}$, or $\hat{\mathbf{v}} \cdot \mathbf{r}$. To turn this quantity into a vector in $\mathbf{v}$-direction, namely $\mathbf{r}_{\parallel}$, one has to multiply it by $\hat{\mathbf{v}}$, i.e. $(\hat{\mathbf{v}} \cdot \mathbf{r})\hat{\mathbf{v}}$. With $\hat{\mathbf{v}} = \mathbf{v}/v$, we obtain

$$\mathbf{r}_{\parallel} = \frac{(\mathbf{v} \cdot \mathbf{r})\mathbf{v}}{v^2} = \frac{(\boldsymbol{\beta} \cdot \mathbf{r})\boldsymbol{\beta}}{\beta^2}. \tag{18.4}$$

For $\mathbf{r}_{\perp}$ follows immediately

$$\mathbf{r}_{\perp} = \mathbf{r} - \frac{(\boldsymbol{\beta} \cdot \mathbf{r})\boldsymbol{\beta}}{\beta^2}. \tag{18.5}$$

Inserted into the Lorentz transformation, we obtain

$$\mathbf{r}' = \mathbf{r}'_{\parallel} + \mathbf{r}'_{\perp} = \gamma \left[\frac{(\boldsymbol{\beta} \cdot \mathbf{r})\boldsymbol{\beta}}{\beta^2} - \boldsymbol{\beta} ct \right] + \mathbf{r} - \frac{(\boldsymbol{\beta} \cdot \mathbf{r})\boldsymbol{\beta}}{\beta^2},$$

or

$$ct' = \gamma(ct - \boldsymbol{\beta} \cdot \mathbf{r})$$
$$\mathbf{r}' = \mathbf{r} + (\gamma - 1)\frac{(\boldsymbol{\beta} \cdot \mathbf{r})\boldsymbol{\beta}}{\beta^2} - \gamma \boldsymbol{\beta} ct$$
$$\gamma = \frac{1}{\sqrt{1 - \beta^2}} = \frac{1}{\sqrt{1 - (\beta_x^2 + \beta_y^2 + \beta_z^2)}}. \tag{18.6}$$

The inverse Lorentz transformation is

$$ct = \gamma(ct' + \boldsymbol{\beta} \cdot \mathbf{r}')$$
$$\mathbf{r} = \mathbf{r}' + (\gamma - 1)\frac{(\boldsymbol{\beta} \cdot \mathbf{r}')\boldsymbol{\beta}}{\beta^2} + \gamma \boldsymbol{\beta} ct'. \tag{18.7}$$

The components of (18.6) are

$$
\begin{pmatrix} ct' \\ x' \\ y' \\ z' \end{pmatrix} = \begin{pmatrix} \gamma & -\gamma\beta_x & -\gamma\beta_y & -\gamma\beta_z \\ -\gamma\beta_x & 1+(\gamma-1)\dfrac{\beta_x^2}{\beta^2} & (\gamma-1)\dfrac{\beta_x\beta_y}{\beta^2} & (\gamma-1)\dfrac{\beta_x\beta_z}{\beta^2} \\ -\gamma\beta_y & (\gamma-1)\dfrac{\beta_y\beta_x}{\beta^2} & 1+(\gamma-1)\dfrac{\beta_y^2}{\beta^2} & (\gamma-1)\dfrac{\beta_y\beta_z}{\beta^2} \\ -\gamma\beta_z & (\gamma-1)\dfrac{\beta_z\beta_x}{\beta^2} & (\gamma-1)\dfrac{\beta_z\beta_y}{\beta^2} & 1+(\gamma-1)\dfrac{\beta_z^2}{\beta^2} \end{pmatrix} \begin{pmatrix} ct \\ x \\ y \\ z \end{pmatrix}
$$

$$(18.8)$$

By setting $\beta_z = 0$, $\beta_x = \beta\cos\delta$, and $\beta_y = \beta\sin\delta$, we re-derive (18.2).

With (18.6), we are now able to *generalize the transformation equations of velocities* (Chapter 9): The S'-frame moves at $\mathbf{v}/c = \boldsymbol{\beta} = (\beta_x, \beta_y, \beta_z)$ relative to the S-frame. In the S'-frame an object moves at $\mathbf{u}' = (u'_x, u'_y, u'_z)$. What is its velocity $\mathbf{u} = (u_x, u_y, u_z)$ in the S-frame? (18.6) yields

$$
\mathbf{u}' = \frac{d\mathbf{r}'}{dt'} = \frac{d\mathbf{r} + (\gamma-1)\dfrac{(\boldsymbol{\beta}\cdot d\mathbf{r})\boldsymbol{\beta}}{\beta^2} - \gamma\boldsymbol{\beta}\,c\,dt}{\gamma\,dt - \dfrac{1}{c}\gamma\boldsymbol{\beta}\cdot d\mathbf{r}},
$$

$$
\mathbf{u}' = \frac{\dfrac{1}{\gamma}\mathbf{u} + \left(1 - \dfrac{1}{\gamma}\right)\dfrac{(\boldsymbol{\beta}\cdot\mathbf{u})\boldsymbol{\beta}}{\beta^2} - \boldsymbol{\beta}\,c}{1 - \dfrac{1}{c}\boldsymbol{\beta}\cdot\mathbf{u}}.
$$

Therefore, the *most general form of the addition theorem of velocities* is

$$
\mathbf{u}' = \frac{1}{1 - \dfrac{\mathbf{u}\cdot\mathbf{v}}{c^2}}\left\{\frac{\mathbf{u}}{\gamma} + \left[\left(1 - \frac{1}{\gamma}\right)\frac{\mathbf{u}\cdot\mathbf{v}}{v^2} - 1\right]\mathbf{v}\right\}
$$

$$
\mathbf{u} = \frac{1}{1 + \dfrac{\mathbf{u}'\cdot\mathbf{v}}{c^2}}\left\{\frac{\mathbf{u}'}{\gamma} - \left[-\left(1 - \frac{1}{\gamma}\right)\frac{\mathbf{u}'\cdot\mathbf{v}}{v^2} - 1\right]\mathbf{v}\right\}. \qquad (18.9)
$$

Chapter 19*

The Rod-Through-The-Window Paradox

The Lorentz transformation in two directions provides us with a tool to quantitatively solve another problem of the special theory of relativity, the *rod-through-the-window paradox*: A rod of proper length ℓ along the x-direction and infinitesimally thin in the y-direction is at rest in the S-frame; the rod's left end is the coordinate origin. A frame, hereinafter called window, lying in the x–z-plane with the proper width ℓ along the x-direction, and also infinitesimally thin in the y-direction, is flying toward the rod from the lower left at a velocity of $\mathbf{v}/c = \boldsymbol{\beta} = (\beta_x, \beta_y, 0)$ in such a way that the window's left end passes exactly through the coordinate origin (Fig. 19.1, left-hand side). At non-relativistic speeds, the window's left and right ends come infinitesimally close to the rod's left and right ends. There is *no collision of window and rod* either at the left or at the right point of contact as *both objects are equally long and infinitesimally thin* in the y-direction.

At relativistic velocities, however, one should expect a length contraction of the window and thus a collision with the rod. But if we put ourselves in the rest frame of the window (S'-frame) (Fig. 19.1, right-hand side), one expects a length contraction of the rod and thus the rod's passing through the window without a problem. Obviously, these events are mutually exclusive. We may already suspect that, similar to the paradoxes of Chapter 8, the idea of both ends passing the window *simultaneously* is the problem.

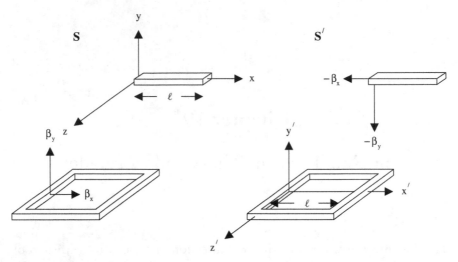

Fig. 19.1. Rod-through-the-window paradox; left: S-frame – the window is moving relative to the rod; right: S'-frame – the rod is moving relative to the window.

Let us first calculate the situation in the S'-frame. We convert the rod's coordinates, measured in the S-frame, into coordinates of the S'-frame:

$$\text{rod's left end in } S \qquad\qquad \text{rod's right end in } S$$
$$\begin{pmatrix} ct_1 \\ x_1 \\ y_1 \end{pmatrix} = \begin{pmatrix} ct_1 \\ 0 \\ 0 \end{pmatrix} ; \qquad\qquad \begin{pmatrix} ct_r \\ x_r \\ y_r \end{pmatrix} = \begin{pmatrix} ct_r \\ \ell \\ 0 \end{pmatrix} .$$

The S'-frame moves relative to the S-frame at $\boldsymbol{\beta} = (\beta_x, \beta_y, 0)$. From (18.8) follows

$$\begin{pmatrix} ct_1' \\ x_1' \\ y_1' \end{pmatrix} = \begin{pmatrix} \gamma ct_1 \\ -\gamma\beta_x ct_1 \\ -\gamma\beta_y ct_1 \end{pmatrix} ; \qquad \begin{pmatrix} ct_r' \\ x_r' \\ y_r' \end{pmatrix} = \begin{pmatrix} \gamma ct_r - \gamma\beta_x \ell \\ -\gamma\beta_x ct_r + \left[1 + (\gamma - 1)\dfrac{\beta_x^2}{\beta^2} \right]\ell \\ -\gamma\beta_y ct_r + (\gamma - 1)\dfrac{\beta_y \beta_x}{\beta^2}\ell \end{pmatrix} .$$

$$\tag{19.1}$$

Now we insert the first components of each of these two column vectors into their second and third components in order to obtain the spatial coordinates as a function of time in the S'-frame:

$$\begin{pmatrix} x_1' \\ y_1' \end{pmatrix} = \begin{pmatrix} -\beta_x ct_1' \\ -\beta_y ct_1' \end{pmatrix} \tag{19.2}$$

$$\begin{pmatrix} x_r' \\ y_r' \end{pmatrix} = \begin{pmatrix} -\beta_x (ct_r' + \gamma \beta_x \ell) + \left[1 + (\gamma - 1)\dfrac{\beta_x^2}{\beta^2}\right]\ell \\[3mm] -\beta_y (ct_r' + \gamma \beta_x \ell) + (\gamma - 1)\dfrac{\beta_y \beta_x}{\beta^2}\ell \end{pmatrix}$$

$$= \begin{pmatrix} -\beta_x ct_r' + \left[1 + (\gamma - 1 - \gamma\beta^2)\dfrac{\beta_x^2}{\beta^2}\right]\ell \\[3mm] -\beta_y ct_r' + (\gamma - 1 - \gamma\beta^2)\dfrac{\beta_y \beta_x}{\beta^2}\ell \end{pmatrix}$$

$$= \begin{pmatrix} -\beta_x ct_r' + \left(1 - \dfrac{\gamma}{\gamma + 1}\beta_x^2\right)\ell \\[3mm] -\beta_y ct_r' - \dfrac{\gamma}{\gamma + 1}\beta_y \beta_x \ell \end{pmatrix}. \tag{19.3}$$

The last conversion was done with

$$\frac{\gamma - 1 - \gamma\beta^2}{\beta^2} = \frac{\gamma(1 - \beta^2) - 1}{\beta^2} = \frac{\dfrac{1}{\gamma} - 1}{1 - \dfrac{1}{\gamma^2}}$$

$$= -\frac{1 - \dfrac{1}{\gamma}}{\left(1 - \dfrac{1}{\gamma}\right)\left(1 + \dfrac{1}{\gamma}\right)} = -\frac{\gamma}{\gamma + 1}.$$

In the S'-frame, the loci of the rod's two ends are to be determined simultaneously, therefore $t_1' = t_r'$. (19.2) can now be inserted into (19.3),

$$\begin{pmatrix} x_r' \\ y_r' \end{pmatrix} = \begin{pmatrix} x_1' + \left(1 - \dfrac{\gamma}{\gamma + 1}\beta_x^2\right)\ell \\[3mm] y_1' - \dfrac{\gamma}{\gamma + 1}\beta_y \beta_x \ell \end{pmatrix}. \tag{19.4}$$

In the S'-frame, the rod's length in the x'-direction is $(1 - \frac{\gamma}{\gamma+1}\beta_x^2)\ell$; so it is contracted. The right end is below the left end by $\frac{\gamma}{\gamma+1}\beta_y\beta_x\ell$. Thus, in the S'-frame, the *rod is tilted* by the angle θ, given by

$$\tan\theta = \frac{\gamma\,\beta_x\beta_y}{1 + \gamma\left(1 - \beta_x^2\right)}, \qquad (19.5)$$

as shown in Fig. 19.2. From (19.4), we can calculate the rod's length in the S'-frame to be

$$\ell'^2 = (x_r' - x_1')^2 + (y_r' - y_1')^2 = \left(1 - \frac{\gamma}{\gamma+1}\beta_x^2\right)^2 \ell^2 + \left(-\frac{\gamma}{\gamma+1}\beta_y\beta_x\right)^2 \ell^2$$

$$\ell' = \ell\sqrt{1 - \frac{2\gamma}{\gamma+1}\beta_x^2 + \left(\frac{\gamma}{\gamma+1}\right)^2 \beta_x^2(\beta_x^2 + \beta_y^2)}.$$

With $(\frac{\gamma}{\gamma+1})^2\beta_x^2(\beta_x^2 + \beta_y^2) = (\frac{\gamma}{\gamma+1})^2\beta_x^2(1 - \frac{1}{\gamma^2}) = \frac{\gamma-1}{\gamma+1}\beta_x^2$, we obtain

$$\ell' = \ell\sqrt{1 - \beta_x^2}.$$

The *rod has shrunk* by the known length contraction factor.

The flight path, or trajectory, of the rod's left end in the S'-frame is calculated by eliminating t_1' in (19.2),

$$y_1' = \frac{\beta_y}{\beta_x}x_1'. \qquad (19.6)$$

As expected, the rod's left end touches the left end of the window as for $x_1' = 0$, $y_1' = 0$ holds. In a corresponding manner, the flight path of the rod's right end can be calculated by eliminating t_r' in (19.3),

$$\frac{1}{\beta_x}\left[x_r' - \left(1 - \frac{\gamma}{\gamma+1}\beta_x^2\right)\ell\right] = \frac{1}{\beta_y}\left[y_r' + \frac{\gamma}{\gamma+1}\beta_y\beta_x\ell\right]$$

$$y_r' = \frac{\beta_y}{\beta_x}\left[x_r' - \left(1 - \frac{\gamma}{\gamma+1}\beta_x^2\right)\ell\right] - \frac{\gamma}{\gamma+1}\beta_y\beta_x\ell$$

$$y_r' = \frac{\beta_y}{\beta_x}x_r' - \frac{\beta_y}{\beta_x}\ell. \qquad (19.7)$$

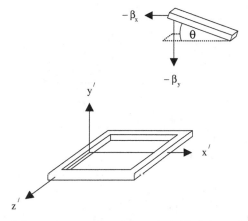

Fig. 19.2. Relativistic correction of the right half of Fig. 19.1: The rod is in motion relative to the window and tilted by the angle θ.

For $y_r' = 0$, $x_r' = \ell$ holds, i.e. the rod's right end touches the right end of the window; *the rod fits perfectly into and through the window!* Fig. 19.2 shows the right half of Fig. 19.1 with the relativistic correction.

The rod touches both ends of the window as it did with non-relativistic velocities, even though its length is reduced by a factor of $\sqrt{1 - \beta_x^2}$. This is possible only because *the rod is tilted in such a way that it requires more space to pass through the window*: The rod's right end passes the right end of the window first, not until after this do the left ends of rod and window pass each other while, in the meantime, the rod's left end has moved to the left.

Let us calculate the situation in the S-frame. First, all window coordinates measured in the S'-frame are transformed into the S-frame:

window's left end in S'

$$\begin{pmatrix} ct_1' \\ x_1' \\ y_1' \end{pmatrix} = \begin{pmatrix} ct_1' \\ 0 \\ 0 \end{pmatrix} ;$$

window's right end in S'

$$\begin{pmatrix} ct_r' \\ x_r' \\ y_r' \end{pmatrix} = \begin{pmatrix} ct_r' \\ \ell \\ 0 \end{pmatrix} .$$

The inverse Lorentz transformation needed here is obtained in the usual way from (18.8), i.e. by interchanging all primed and unprimed variables and all signs of β-factors: only the six minus signs in the first line and the first column of (18.8) turn into plus signs. Thus we obtain from (19.1),

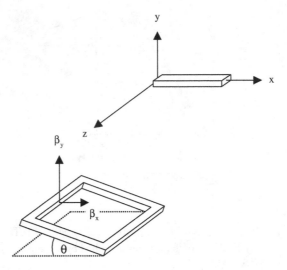

Fig. 19.3.   Relativistic correction of the left half of Fig. 19.1: The window is
in motion relative to the rod and tilted by the angle θ.

$$\begin{pmatrix} ct_1 \\ x_1 \\ y_1 \end{pmatrix} = \begin{pmatrix} \gamma ct_1' \\ \gamma \beta_x ct_1' \\ \gamma \beta_y ct_1' \end{pmatrix} ; \quad \begin{pmatrix} ct_r \\ x_r \\ y_r \end{pmatrix} = \begin{pmatrix} \gamma ct_r' + \gamma \beta_x \ell \\ \gamma \beta_x ct_r' + \left[1 + (\gamma - 1)\dfrac{\beta_x^2}{\beta^2} \right] \ell \\ \gamma \beta_y ct_r' + (\gamma - 1)\dfrac{\beta_y \beta_x}{\beta^2} \ell \end{pmatrix} .$$

$$(19.8)$$

The calculation continues by analogy with the above derivation, and a
result similar to (19.4) follows, except with unprimed variables. In the
x-direction, the window is contracted by a factor of $1 - \frac{\gamma}{\gamma+1}\beta_x^2$, the window's
right end is lower than its left end by $\frac{\gamma}{\gamma+1}\beta_y\beta_x\ell$. Both left ends touch, and
so do the right ends. The window, though having become too narrow,
passes over the rod because *the window is tilted in such a way that the
window needs less space to move over the rod.* The window's left end passes
the left end of the rod first, not until after this do the right ends of rod and
window pass each other while, in the meantime, the window's right end
has moved to the right. Fig. 19.3 shows the left half of Fig. 19.1 with the
relativistic correction.

We have now arrived at a quantitative solution for the rod-through-the-window paradox with the help of a Lorentz transformation in two directions and found that not only does the rod fit perfectly into the window, but the window also passes perfectly over the rod. What still seems paradoxical is that the parallelism of rod and window is lost. A clever trick promotes our understanding that this *must* be so and that, basically, the whole problem has already been dealt with at the end of Chapter 4.

For this purpose, we introduce x^*- and y^*-coordinates. x^* is aligned with the relative motion of rod and window (Fig. 19.4, top). In the S-frame (rod at rest, window in motion), the window is subject to a length contraction in the x^*-direction, but not in the y^*-direction (Fig. 19.4, center). This is the reason for the window to move over the rod without a problem. But in so doing, the window straightens up and is no longer

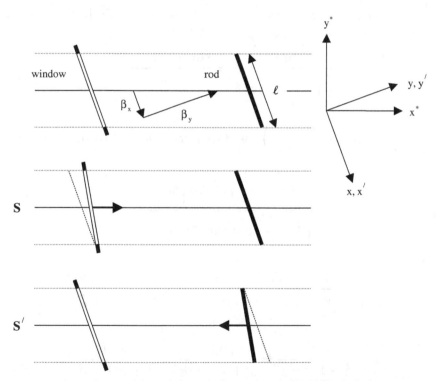

Fig. 19.4. Solution of the rod-through-the-window paradox by introducing a new coordinate frame (top); center: window is moving relative to the rod at rest; bottom: rod is moving relative to the window at rest.

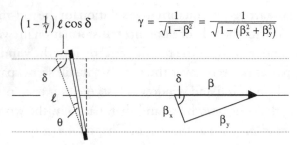

Fig. 19.5.   Quantitative treatment of Fig. 19.4, center. θ is the tilt of the window of proper length ℓ, δ the angle between the direction of motion and the x-axis.

parallel to the rod. In the S'-frame (window at rest, rod in motion), the rod is subject to a length contraction in the x^*-direction (Fig. 19.4, bottom), the rod straightens up and fits perfectly into the window.

We can also check these considerations quantitatively (Fig. 19.5). We write down the sine theorem,

$$\frac{\sin\theta}{\left(1 - \frac{1}{\gamma}\right)\ell\cos\delta} = \frac{\sin[\pi - (\delta + \theta)]}{\ell}$$

$$\sin\theta = \left(1 - \frac{1}{\gamma}\right)\cos\delta(\sin\delta\cos\theta + \cos\delta\sin\theta)$$

$$\frac{\left[1 - \left(1 - \frac{1}{\gamma}\right)\cos^2\delta\right]\sin\theta}{\left(1 - \frac{1}{\gamma}\right)\cos\delta\sin\delta\cos\theta} = 1$$

$$\tan\theta = \frac{(\gamma - 1)\cos\delta\sin\delta}{\gamma - (\gamma - 1)\cos^2\delta} = \frac{(\gamma - 1)\beta_x\beta_y}{\gamma\beta^2 - (\gamma - 1)\beta_x^2} = \frac{\gamma\,\beta_x\beta_y}{\dfrac{\gamma^2\beta^2}{\gamma - 1} - \gamma\beta_x^2}$$

$$= \frac{\gamma\beta_x\beta_y}{\dfrac{\gamma^2 - 1}{\gamma - 1} - \gamma\beta_x^2} = \frac{\gamma\beta_x\beta_y}{\gamma + 1 - \gamma\beta_x^2} = \frac{\gamma\beta_x\beta_y}{1 + \gamma(1 - \beta_x^2)}.$$

This is (19.5), Q.E.D.

Problems for Part II

1. **Addition of velocities I**
   Spaceship A moves away from earth at a velocity of $u_{xA} = -0.50c$, spaceship B at $u_{xB} = 0.70c$.

   (a) How fast is B's velocity in A's reference frame?
   (b) How does it change when the components $u_{yB} = -0.20c$ and $u_{zB} = 0.30c$ are added?

2. **Addition of velocities II**
   $u'^2 = u_x'^2 + u_y'^2 + u_z'^2$, $u^2 = u_x^2 + u_y^2 + u_z^2$, and $\mathbf{v} = (v, 0, 0)$ are given.
   Show that $1 - \frac{u'^2}{c^2} = (1 - \frac{u'^2}{c^2})(1 - \frac{v^2}{c^2})(1 + \frac{u_x' v}{c})^{-2}$ is valid. Then use this equation to show that the sum of two velocities that are both less than c is always less than c.

3. **Aberration of light**
   Find out what the parallax of a star is.

   (a) Draw a sketch of the earth's orbit around the sun and show for a star in the ecliptic plane that the parallactic motion and the motion of aberration are 90° out of phase.
   (b) How close to earth (in km and ly) would a star have to be for his parallax and his aberration to be of equal size?

4. **Search light effect**

   Give a rough estimate by what factor the intensity of a fast moving radiation source ($\beta = 0.9; 0.99; 0.999$) increases (or decreases) into (or opposite to) the direction of motion. Use the density of rays, as in Fig. 10.2, as a measure of intensity and take into account that the drawing is only two-dimensional. Example: In Fig. 10.2, the density of rays in the direction of motion increases by about a factor of 5, therefore the intensity is higher by about a factor of 25. Finally, solve the problem analytically by calculating $(d\theta'/d\theta)^2_{\theta=0°;180°}$.

5. **Transformation of accelerations**

   Spaceship B in Problem 1b is subject to an acceleration of the size $a_x = a_y = a_z = 10\,\text{m/s}^2$ in the earth's reference frame. What is its acceleration in the frame of spaceship A?

6. **Higher mathematics**

   (a) Insert (12.12) into (12.11) and rearrange until you find $\tau = \tau$.
   (b) Differentiate (12.7) in order to obtain (12.6) again.
   (c) Treat (12.8) correspondingly.
   (d) Derive (12.14) from (12.13) and the inverse function of (12.12) by writing $u_x = \frac{dx}{dt} = \frac{dx}{d\tau}\frac{d\tau}{dt}$.

7. **Rocket-rope paradox**

   (a) Insert (12.14) into (13.10) and thus calculate the distance between A and B in the A-frame as a function of the proper time of A.
   (b) Calculate the velocity of B in the A-frame by taking the derivative w.r.t. the proper time.
   (c) Where do the results of (a) and (b) tend to for $\tau \to \infty$? Note that A is not an inertial frame and use Fig. 13.1 to understand the results.

8. **Doppler effect and time dilation**

   A spaceship is moving at $\beta = +0.50; -0.50$ relative to earth and is emitting light signals to earth with a period of $1.00\,\text{s}$. Use (14.3) and (2.1), respectively, to calculate the period on earth. Why are the results different, and what do they mean?

9. **Doppler effect and accelerated motion**

   An astronaut is moving away from earth at a proper acceleration of $\alpha = 10\,\text{m/s}^2$ while emitting radiation. After how much time and how much proper time, respectively, has the frequency of the radiation received on earth halved?

10. **Triplet paradox**

    Conditions are as in Chapter 15, except that triplets are given: Triplet C spends 10 years at $\beta = -0.8$ to fly to a star 8 light-years to the left of earth and spends another 10 years to return to earth at $\beta = 0.8$. Triplets B and C leave earth on 1.1.2000.

    Draw in one Minkowski diagram, like Fig. 15.1, five pairs of axes: (t_A, x_A), as well as (t'_B, x'_B) and (t''_B, x''_B) for B's journey to and from the star, and correspondingly (t'_C, x'_C) and (t''_C, x''_C). Show graphically that B makes a time-leap in C's frame and starts returning before arrival on his star.

11. **Image of a fast moving UFO**

    A UFO passes earth on a straight line at a velocity of $\beta = 0.8$. At the time $t = 0\,\text{s}$, it approaches earth as close as 4 light-seconds.

    (a) Calculate its actual and its apparent positions for $t = -3\,\text{s}$, $-2\,\text{s}, \ldots, 6\,\text{s}$.

    (b) Calculate the stretching or shrinking factor for a very short UFO. Hint: Make a Taylor expansion of the corresponding formulas for $\ell \to 0$.

    (c) Calculate the UFO's apparent velocities for the times given above.

12. **Lorentz transformation in two spatial directions**

    The S'-frame moves relative to the S-frame at a velocity of $\mathbf{v} = (c/2, c/2, 0)$. Calculate the Lorentz transformation correctly according to (18.8) and incorrectly by applying two Lorentz transformations in sequence: first one into the x-direction, followed by another one into the y-direction, and also in reversed order.

PART III
Dynamics

Chapter 20

Mass and Momentum

The S'-frame moves at an arbitrary but constant velocity $v < c$ parallel to the x-axis of the S-frame, with both frames having mutually parallel axes (Fig. 20.1). In the S'-frame an object A of mass m' moves at a low, non-relativistic velocity u'_y parallel to the y'-axis. From the point of view of an observer B at rest in the S-frame, A moves at a velocity of v parallel to the x-axis and, with (9.1), at $u_y = u'_y/\gamma = u'_y\sqrt{1 - (\frac{v}{c})^2}$ parallel to the y-axis.

What can we say about the momentum of A in the S-frame and in the S'-frame? In the x-direction, i.e. the direction of motion of S' relative to S, A's momentum vanishes in the S'-frame, but does not vanish in the S-frame, except for the case of $v = 0$. Therefore, *in the longitudinal direction*, i.e. in the direction parallel to the relative motion of the two frames, *momentum always has to be transformed* when one changes from one inertial frame to another one.

In the transverse direction, i.e. in a direction orthogonal to the relative motion of the two frames, $u_y = u'_y$ holds for non-relativistic relative velocities v. However, for relativistic relative velocities (9.1) holds, which

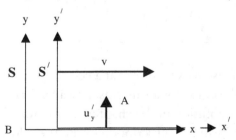

Fig. 20.1. Derivation of the velocity dependence of mass.

177

implies a need for transforming momentum in the transverse direction as well.

We shall see in Part IV that such a procedure would be very cumbersome: There would no longer be a Lorentz transformation for the momentum vector, which would turn momentum into a rather troublesome quantity for relativistic calculations. For this reason *we postulate that the transverse momentum shall be the same in both inertial reference frames.* This postulate, however, has far-reaching consequences for the mass of the object in motion because we have to write

$$p_y = p'_y$$
$$mu_y = m'u'_y$$
$$mu'_y\sqrt{1 - \left(\frac{v}{c}\right)^2} = m'u'_y$$
$$m = \frac{m'}{\sqrt{1 - \left(\frac{v}{c}\right)^2}}.$$

m' is the mass of the object in the S'-frame, i.e. in the frame where the object has only a low, non-relativistic velocity u'_y. It is thus called the *rest mass* and will from now on be denoted by m_0. m is the same mass measured in the S-frame. Obviously, it is greater than the rest mass; its size depends on the relative velocity v between the S- and the S'-frame. It is thus called the *relativistic mass*. We replace v by u and write

$$m = \frac{m_0}{\sqrt{1 - \left(\frac{u}{c}\right)^2}} = \frac{m_0}{\sqrt{1 - \beta_u^2}} = \gamma_u m_0. \qquad (20.1)$$

If an object of rest mass m_0 moves at a velocity u relative to an observer, he will find that the object's mass increases by a factor of γ_u to $m = \gamma_u m_0$. In order to avoid confusion in the chapters to follow, all β- and γ-factors will from now on include a subscript denoting the velocity which appears under the square root term.

The derivation of the above equation still requires one explanation: In the second line a low, i.e. *non-relativistic*, transverse velocity u'_y is used. u'_y has to be non-relativistic at this point. If we chose a relativistic velocity instead, the relativistic mass equation (20.1), which is to be derived from this calculation, would already have to be applied on the right-hand side. But this procedure does not incur a loss of generality, as we can see by redoing the whole calculation with a relativistic u'_y. In the S'-frame, A's mass would then increase by a factor of $(1 - (\frac{u'_y}{c})^2)^{-1/2}$, and we would have to write

$$ mu_y = \frac{m' u'_y}{\sqrt{1 - \left(\dfrac{u'_y}{c}\right)^2}} = \frac{m' u_y}{\sqrt{1 - \left(\dfrac{u'_y}{c}\right)^2} \sqrt{1 - \left(\dfrac{v}{c}\right)^2}} $$

$$ m = \frac{m'}{\sqrt{1 - \dfrac{\left(\dfrac{u_y}{c}\right)^2}{1 - \left(\dfrac{v}{c}\right)^2}} \sqrt{1 - \left(\dfrac{v}{c}\right)^2}} = \frac{m'}{\sqrt{1 - \dfrac{v^2 + u_y^2}{c^2}}}. $$

For the transformation from u'_y to u_y, (9.1) was used. $v^2 + u_y^2$ is the square of the magnitude of the velocity of A in the S-frame. The mass exhibits the same dependence on velocity as in (20.1) where u_y^2 did not appear because of its negligible size.

In Chapters 2 and 4, the terms *proper time* and *proper length* were introduced, now the notion of a *rest mass* is added. Let us summarize: When the duration of a process, a length, or a mass is measured in an inertial reference frame where this process, this length, or this mass is at rest, one measures a shorter time, a longer length, or a smaller mass than in any other inertial reference frame. After *relativization of time and length* (or space), *relativization of mass* now appears, thus completing the fundamental system of units.

The ratio of relativistic mass to rest mass, rounded to four digits, is:

β_u	0.0	0.1	0.2	0.8	0.9	0.99	0.999	1.0
m/m_0	1.000	1.005	1.021	1.667	2.294	7.089	22.37	∞

Upon approaching the speed of light, this mass ratio increases beyond all bounds. An infinite amount of energy would be required to accelerate an object to c. In other words: *no material object with a finite rest mass can ever be accelerated to the speed of light*, not to mention beyond. This is one of the most crucial insights of the special theory of relativity.

With the velocity dependence of mass we can immediately write down the relativistic momentum $\mathbf{p}$ of an object of rest mass m_0, moving at the velocity $\mathbf{u}$ relative to an observer,

$$\mathbf{p} = m\mathbf{u} = \frac{m_0\mathbf{u}}{\sqrt{1-\left(\dfrac{u}{c}\right)^2}} = \frac{m_0\mathbf{u}}{\sqrt{1-\beta_u^2}} = \gamma_u m_0\mathbf{u}. \qquad (20.2)$$

(20.1) can also be derived by a thought experiment which makes use of the idea that the location of the center of mass of a system of masses does not depend on the inertial reference frame chosen to describe it. Imagine a launch pad with two spaceships at rest, A and B, both having the same rest mass, m_0. The spaceships are catapulted apart by an elastic spring giving them the constant velocities $-u$ and $+u$, respectively, in the S-frame of the launch pad. The center of mass of the system remains on the launch pad (Fig. 20.2).

S-frame

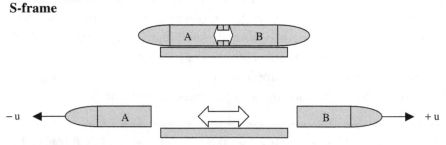

Fig. 20.2.   Derivation of the velocity dependence of mass from a thought experiment with a pair of spaceships catapulted into opposite directions from a common launch pad.

S′-frame

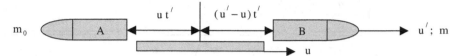

Fig. 20.3. The system of Fig 20.2 in the S'-frame where spaceship A is at rest.

After launch, spaceship A shall be the origin of the S'-frame. In this frame, the launch pad has the velocity u and, according to (9.1), spaceship B has the velocity

$$u' = \frac{2u}{1 + \dfrac{u^2}{c^2}} < 2u. \qquad (20.3)$$

Figure 20.3 shows the system of two spaceships in the S'-frame.

We now demand that the center of mass of the system remains on the launch pad in the S'-frame as well. A new mass, $m \neq m_0$, has to be assigned to spaceship B, yielding

$$m_0 u t' = m(u' - u)t', \qquad (20.4)$$

with ut' and $(u' - u)t'$, respectively, being the distances of the spaceships from the launch pad. The addition theorem of velocities does not have to be applied on the right-hand side as both u' and u are velocities of the S'-frame. First, we take (20.3) to turn $u' - u$ into a function of u and obtain

$$u' - u = \frac{2u}{1 + \dfrac{u^2}{c^2}} - u = \frac{2uc^2}{c^2 + u^2} - \frac{u^3 + uc^2}{c^2 + u^2} = \frac{u(c^2 - u^2)}{c^2 + u^2}. \qquad (20.5)$$

This turns (20.4) into

$$m = m_0 \frac{c^2 + u^2}{c^2 - u^2}.$$

Now u has to be expressed by u'. To do so we rearrange terms,

$$m = m_0 \frac{1 + \dfrac{u^2}{c^2}}{\sqrt{\left(1 - \dfrac{u^2}{c^2}\right)^2}} = m_0 \frac{1 + \dfrac{u^2}{c^2}}{\sqrt{\left(1 + \dfrac{u^2}{c^2}\right)^2 - 4\dfrac{u^2}{c^2}}} = \frac{m_0}{\sqrt{1 - \dfrac{4\dfrac{u^2}{c^2}}{\left(1 + \dfrac{u^2}{c^2}\right)^2}}}.$$

With (20.3), we obtain

$$m = \frac{m_0}{\sqrt{1 - \left(\dfrac{u'}{c}\right)^2}},$$

Q.E.D.

Chapter 21*

Motion of a Rocket

The problem is the following: A rocket of initial mass M_i and initial velocity $u_i = 0$ is launched in a space of zero gravity. It continuously ejects a mass of propellant gas to the rear with the velocity u_0 until the rocket's mass is reduced to a final value of M_f. What is the final velocity u_f of the rocket?

Let us *first calculate non-relativistically*: At any moment during this accelerated motion, the rocket has a velocity of u and a mass of M. After ejection of an infinitesimal amount of gas of mass dm at a velocity of magnitude u_0 relative to the rocket, the rocket has a mass of $M - dm$ and a velocity of $u + du$ (Fig. 21.1).

In the laboratory frame, or rest frame of the launch pad, the rocket's momentum before ejecting the infinitesimal amount of gas dm equals Mu; after ejection, the rocket's momentum is $(M - dm)(u + du)$, and the momentum of the gas is $dm(u - u_0)$. Since there is no external force on rocket and gas, the momentum of the rocket-gas system remains constant,

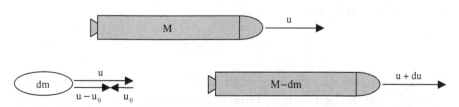

Fig. 21.1. Rocket of mass M in the laboratory frame ejecting gas of mass dm at the velocity u_0.

$$Mu = Mu + M\,du - u\,dm - dm\,du + u\,dm - u_0\,dm. \qquad (21.1)$$

This equation contains three types of products: (1) *products of two variables neither of which is infinitesimal*, namely Mu; (2) *products of two variables one of which is infinitesimal*, namely $M\,du$, $u\,dm$, and $u_0\,dm$; and (3) a *product of two infinitesimal quantities*, $dm\,du$. Products of the second type are *infinitesimal*, i.e. *infinitely much smaller* than products of the first type. Products of the third type are, once again, *infinitely much smaller* than products of the second type. If an equation contains products of the first type, products of the second and third types *can be neglected without incurring the slightest error*. Since both products of the first type cancel out, the products of the second type are the largest remaining summands; therefore the product of the third type can be neglected. Neglecting such products should be distinguished from neglecting higher order terms in many of the Taylor expansions used so far: Those higher order terms were *much smaller*, but not *infinitely much smaller* than their predecessors; their neglect gave rise to smaller or larger numerical errors; thus results obtained in that way were *only approximations*.

In (21.1), two products remain which are both greater than zero and form the simple differential equation

$$M\,du = u_0\,dm. \qquad (21.2)$$

This equation is nothing but the momentum conservation law after ejection of gas of the mass dm in the instantaneous rest frame of the rocket before ejection of that mass: $M\,du$ is the momentum of the rocket (going right), $u_0\,dm$ is the momentum of the gas (going left). What is left to be considered is that the mass increase of the ejected gas, $+dm$, equals the mass decrease of the rocket, $-dM$, i.e.

$$M\,du = -u_0\,dM. \qquad (21.3)$$

This equation, without the foregoing introductory words, can be found in many books as the starting line of the rocket problem. The lengthy remarks in this book are meant to provide an additional, and hopefully valuable, exercise in differential calculus. (21.3) can be integrated in a straightforward manner,

$$\int_0^{u_f} du = -u_0 \int_{M_i}^{M_f} \frac{dM}{M},$$

to give

$$u_f = u_0 \ln \frac{M_i}{M_f}. \tag{21.4}$$

This equation is known as *Ziolkowski's rocket equation.* It is the non-relativistic solution of the differential equation (21.3). As a practical example for (21.4), one might calculate what fraction of the mass of a conventional one-stage rocket reaches the earth's orbit. When typical values are inserted, such as $u_f = 8 \, km/s$ and $u_0 = 4 \, km/s$, M_f/M_i turns out to be 0.14; 14% of the initial mass reaches the orbit, 86% is fuel.

Where does the *relativistic treatment* of the rocket problem deviate from the above derivation? An observer in the laboratory frame writes (21.3) with primes because all variables are measured in the S'-frame of the rocket,

$$M' \, du' = -u_0' \, dM'. \tag{21.5}$$

du', the velocity increase of the rocket with a rest mass of M' in its instantaneous rest frame, is *not* equal to its velocity increase measured in the laboratory frame, du. Therefore, the addition theorem of velocities, (9.1), has to be invoked to find

$$u + du = \frac{u + du'}{1 + \frac{u du'}{c^2}}. \tag{21.6}$$

u is the instantaneous velocity of the rocket in the laboratory frame. To reiterate, a clear distinction has to be made between the laboratory frame (unprimed variables) and the instantaneous rest frame (primed variables). (21.6) is then subjected to a Taylor expansion, giving

$$u + du = (u + du') \left(1 - \frac{u du'}{c^2} \right). \tag{21.7}$$

The Taylor expansion, aborted after the first order term, is – similar to the one in Chapter 16 – *not an approximation but an exact conversion* as $u \, du'$ is infinitely much smaller than c^2, i.e. $u \, du'/c^2$ is infinitely much smaller than unity. (21.7) can be transformed further to read

$$u + du = u + du' \left(1 - \frac{u^2}{c^2}\right).$$ (21.8)

Again, this is *not an approximation but an exact conversion* as only the product of two infinitesimal quantities, $du' \, du'$, is neglected. Thus,

$$du' = \frac{du}{1 - \left(\dfrac{u}{c}\right)^2} = \gamma_u^2 \, du.$$ (21.9)

This turns (21.5) into

$$\int_0^{u_f} \frac{du}{1 - \left(\dfrac{u}{c}\right)^2} = -u_0' \int_{M_i'}^{M_f'} \frac{dM'}{M'},$$ (21.10)

from which follows

$$\frac{c}{2} \ln \frac{1 + \dfrac{u_f}{c}}{1 - \dfrac{u_f}{c}} = u_0' \ln \frac{M_i'}{M_f'},$$ (21.11)

as can be shown by differentiation. For the sake of simplicity, we will *hereafter omit all primes*, but keep in mind that M_i and M_f are rest masses, i.e. masses measured in the reference frame of the rocket, and that the propellant velocity, u_0, is measured in the reference frame of the rocket as well.

We first want to prove that this result is *consistent with the non-relativistic result* for the limiting case of $u_f/c \ll 1$. With three Taylor expansions in series one arrives via

$$\frac{1 + \dfrac{u_f}{c}}{1 - \dfrac{u_f}{c}} \approx \left(1 + \frac{u_f}{c}\right)\left(1 + \frac{u_f}{c}\right) \approx 1 + 2\frac{u_f}{c}$$

and

$$\ln\left(1 + 2\frac{u_f}{c}\right) \approx 2\frac{u_f}{c}$$

at the known result, (21.4), Q.E.D.

One now has to solve (21.11) for u_f which is reached after a couple of elementary transformations:

$$\ln\left(\frac{1 + \dfrac{u_f}{c}}{1 - \dfrac{u_f}{c}}\right)^{\frac{c}{2}} = \ln\left(\frac{M_i}{M_f}\right)^{u_0}$$

$$\frac{1 + \dfrac{u_f}{c}}{1 - \dfrac{u_f}{c}} = \left(\frac{M_i}{M_f}\right)^{\frac{2u_0}{c}}$$

$$\left(1 + \frac{u_f}{c}\right)\left(\frac{M_f}{M_i}\right)^{\frac{2u_0}{c}} = 1 - \frac{u_f}{c}$$

$$\frac{u_f}{c}\left[\left(\frac{M_f}{M_i}\right)^{\frac{2u_0}{c}} + 1\right] = 1 - \left(\frac{M_f}{M_i}\right)^{\frac{2u_0}{c}}$$

$$u_f = c\,\frac{1 - \left(\dfrac{M_f}{M_i}\right)^{\frac{2u_0}{c}}}{1 + \left(\dfrac{M_f}{M_i}\right)^{\frac{2u_0}{c}}}. \tag{21.12}$$

This is the *relativistic generalization of Ziolkowski's rocket equation*. The most efficient rocket is a photon rocket because its fuel are photons which are emitted at the highest velocity possible, namely $u_0 = c$. Then the relativistic Ziolkowski rocket equation simply reads

$$u_f = c \frac{1 - \left(\dfrac{M_f}{M_i}\right)^2}{1 + \left(\dfrac{M_f}{M_i}\right)^2}. \tag{21.13}$$

Let us now take (12.14) in order to express the final velocity as a function of the proper time τ elapsed in the rocket and of the proper acceleration α,

$$u_f = c \tanh\left(\frac{\alpha\tau}{c}\right),$$

and equate this with (21.12) to obtain

$$\frac{1 - \left(\dfrac{M_f}{M_i}\right)^{\frac{2u_0}{c}}}{1 + \left(\dfrac{M_f}{M_i}\right)^{\frac{2u_0}{c}}} = \tanh\left(\frac{\alpha\tau}{c}\right) = \frac{\exp\left(\dfrac{\alpha\tau}{c}\right) - \exp\left(-\dfrac{\alpha\tau}{c}\right)}{\exp\left(\dfrac{\alpha\tau}{c}\right) + \exp\left(-\dfrac{\alpha\tau}{c}\right)}$$

$$= \frac{1 - \exp\left(-\dfrac{2\alpha\tau}{c}\right)}{1 + \exp\left(-\dfrac{2\alpha\tau}{c}\right)}.$$

Comparing left-hand and right-hand sides gives

$$\left(\frac{M_f}{M_i}\right)^{\frac{u_0}{c}} = \exp\left(-\frac{\alpha\tau}{c}\right). \tag{21.14}$$

For a photon rocket, (21.14) is reduced to

$$\frac{M_f}{M_i} = \exp\left(-\frac{\alpha\tau}{c}\right). \tag{21.15}$$

As an example, we consider a space flight comprising four phases:

- acceleration away from earth at $\alpha = g$,
- deceleration upon approaching the destination at $|\alpha| = g$,
- acceleration toward earth, as in the first phase,
- deceleration upon approaching earth, as in the second phase.

All four phases require the same amount of proper time, τ. This gives for the mass ratio

$$\frac{M_f}{M_i} = \exp\left(-\frac{4\alpha\tau}{c}\right)$$

and, with (12.13), for the penetration depth into space

$$2x = \frac{2c^2}{\alpha}\left[\cosh\left(\frac{\alpha\tau}{c}\right) - 1\right]. \qquad (21.16)$$

For $\alpha = g = 10\,\text{m/s}^2$, the following numerical values are obtained:

duration of acceleration, τ (yrs)	duration of flight, 4τ (yrs)	maximum distance, $2x$ (ly)	duration of flight, as seen on earth, t (yrs)	mass ratio M_f/M_i
1	4	1.2	4.8	1.5×10^{-2}
2	8	6.0	15	2.2×10^{-4}
5	20	1.8×10^2	3.6×10^2	7.4×10^{-10}
10	40	3.5×10^4	7.0×10^4	5.5×10^{-19}
20	80	1.3×10^9	2.6×10^9	3.0×10^{-37}

At an acceleration of $10\,\text{m/s}^2$, optimally suited for humans, the visible universe can be traversed in one lifetime. However, the astronauts would return after billions of years, as measured on earth. What makes such space travel utterly impossible in the end is the very unfavorable mass ratio, i.e. the size of the fuel tanks required at the beginning of the journey.

Chapter 22*

Force

In non-relativistic physics, Newton's second law reads *force equals mass times acceleration*. It can also be written as the first time derivative of momentum,

$$\mathbf{f} = m\mathbf{a} = m\frac{d\mathbf{u}}{dt} = \frac{d(m\mathbf{u})}{dt} = \frac{d\mathbf{p}}{dt}. \tag{22.1}$$

Since mass, m, is a constant, it may be written in front of or behind the differential operator. Contrary to the common notation, a lower case variable was chosen for force in order to reserve uppercase variables for four-vectors.

In relativistic physics, mass is no longer a constant; the conversions of line (22.1) are no longer applicable. Force is rather *defined* by

$$\mathbf{f} = \frac{d\mathbf{p}}{dt}. \tag{22.2}$$

Thus, $\mathbf{f} = m\mathbf{a}$ is only a non-relativistic approximation. In relativistic physics, a more complicated relation exists between force $\mathbf{f}$ and acceleration $\mathbf{a}$. To arrive there, an auxiliary calculation is required first,

$$\frac{d\gamma_u}{dt} = \frac{d}{dt}\left[1 - \frac{1}{c^2}(u_x^2 + u_y^2 + u_z^2)\right]^{-1/2}$$

$$= -\frac{1}{2}\left[1 - \frac{1}{c^2}(u_x^2 + u_y^2 + u_z^2)\right]^{-3/2}\left[-\frac{2}{c^2}\left(u_x\frac{du_x}{dt} + u_y\frac{du_y}{dt} + u_z\frac{du_z}{dt}\right)\right]$$

$$\frac{d\gamma_u}{dt} = \frac{\gamma_u^3}{c^2}\mathbf{u}\cdot\frac{d\mathbf{u}}{dt} = \frac{\gamma_u^3}{c^2}\mathbf{u}\cdot\mathbf{a}. \tag{22.3}$$

With (20.2), (22.2) turns into $\mathbf{f} = \frac{d}{dt}(\gamma_u m_0 \mathbf{u})$ and, with (22.3), into

$$
\mathbf{f} = \gamma_u m_0 \mathbf{a} + \gamma_u^3 m_0 \frac{1}{c^2}(\mathbf{u} \cdot \mathbf{a})\mathbf{u}
$$

$$
\gamma_u = \frac{1}{\sqrt{1 - \beta_u^2}} = \frac{1}{\sqrt{1 - \frac{1}{c^2}(u_x^2 + u_y^2 + u_z^2)}}. \qquad (22.4)
$$

This equation is the *relativistic law of motion*. It is interesting to note that both acceleration and velocity enter it and that force and acceleration are no longer colinear. In the non-relativistic case, $\mathbf{u}/c$ approaches zero, and γ_u approaches unity. This reduces (22.4) to the well known equation $\mathbf{f} = m\mathbf{a}$ where $\mathbf{f}$ and $\mathbf{a}$ are colinear.

When it is possible to span a coordinate system such that $\mathbf{u} = (u, 0, 0)$, then (22.4) becomes

$$
f_x = \gamma_u m_0 a_x + \gamma_u^3 m_0 \frac{1}{c^2} u a_x u = \gamma_u m_0 a_x \left(1 + \gamma_u^2 \frac{u^2}{c^2}\right)
$$

$$
= \gamma_u m_0 a_x \left(1 + \frac{\beta_u^2}{1 - \beta_u^2}\right)
$$

$$
f_x = \gamma_u^3 m_0 a_x
$$
$$
f_y = \gamma_u m_0 a_y
$$
$$
f_z = \gamma_u m_0 a_z
$$
$$
\gamma_u = \frac{1}{\sqrt{1 - \beta_u^2}} = \frac{1}{\sqrt{1 - \left(\frac{u}{c}\right)^2}}. \qquad (22.5)
$$

(22.5) can also be interpreted in terms of *force equals mass times acceleration*. Then, however, one has to distinguish between a *longitudinal mass*, $\gamma_u^3 m_0$, for the force parallel to the direction of motion, and a *transverse mass*, $\gamma_u m_0$, for the force perpendicular to the direction of motion.

While (22.1) can immediately be solved for **a**, it is more tricky to do so with (22.4). One starts by forming a scalar product of (22.4) and **u**,

$$\mathbf{f} \cdot \mathbf{u} = \gamma_u m_0 \mathbf{u} \cdot \mathbf{a} \left(1 + \frac{\gamma_u^2}{c^2} \mathbf{u} \cdot \mathbf{u} \right) = \gamma_u m_0 \mathbf{u} \cdot \mathbf{a} \left(1 + \frac{\beta_u^2}{1 - \beta_u^2} \right) = \gamma_u^3 m_0 \mathbf{u} \cdot \mathbf{a}.$$

When this result is inserted into the second summand of (22.4), one obtains

$$\mathbf{f} = \gamma_u m_0 \mathbf{a} + \mathbf{f} \cdot \mathbf{u} \frac{1}{c^2} \mathbf{u},$$

and solves for **a**,

$$\mathbf{a} = \frac{1}{\gamma_u m_0} \left(\mathbf{f} - \frac{\mathbf{f} \cdot \mathbf{u}}{c^2} \mathbf{u} \right). \tag{22.6}$$

As an example of relativistic forces and momenta, let us examine the motion of a particle of rest mass m_0 and charge q in a homogeneous electric field **E**.

Case 1: The *positive charge is at rest at the origin* of the S-frame. At $t = 0$, an electric field E_x is turned on which exerts an electric force f_x on the charge and gives it the momentum $p_x(t)$. We would now like to calculate the velocity $u_x(t)$ of the charge. Of course, p_y is and remains zero at all times (Fig. 22.1).

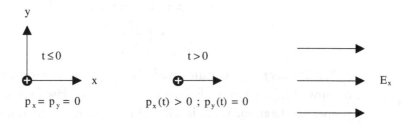

Fig. 22.1. A charge is at rest at the origin at $t \leq 0$, and is accelerated in the x-direction at $t > 0$.

As will be shown in Chapter 34, the force in an electric field is given in *both* non-relativistic *and* relativistic physics by

$$f_x = \frac{dp_x}{dt} = qE_x, \tag{22.7}$$

and integrated,

$$\int_0^{p_x} dp_x = qE_x \int_0^t dt$$

$$p_x = qE_x t. \tag{22.8}$$

With (20.2), i.e.

$$p_x = \frac{m_0 u_x}{\sqrt{1 - \left(\frac{u_x}{c}\right)^2}},$$

(22.8) becomes

$$\frac{m_0 u_x}{\sqrt{1 - \left(\frac{u_x}{c}\right)^2}} = qE_x t$$

$$\left(\frac{u_x}{c}\right)^2 = \left(\frac{qE_x t}{m_0 c}\right)^2 \left[1 - \left(\frac{u_x}{c}\right)^2\right]$$

$$\frac{u_x}{c} = \frac{\dfrac{qE_x t}{m_0 c}}{\sqrt{1 + \left(\dfrac{qE_x t}{m_0 c}\right)^2}}. \tag{22.9}$$

When qE_x/m_0 is identified with the proper acceleration α, one obtains (12.7), i.e. the relativistic velocity-time law for constant proper acceleration. What is still left to be shown is that, indeed,

$$\alpha = \frac{qE_x}{m_0} \tag{22.10}$$

is true for the charge. To do so we remember (12.5), $\alpha = \gamma_{u_x}^3 a_x$, so (22.10) becomes

$$\gamma_{u_x}^3 a_x = \frac{qE_x}{m_0}.$$

With (22.5) we obtain (22.7), or $f_x = qE_x$, Q.E.D.

Case 2: The *positive charge is not at rest at the origin* of the S-frame at $t = 0$, but has a momentum p_0 in the y-direction. At $t = 0$, the electric field E_x is turned on. We would like to calculate the velocity components $u_x(t)$ and $u_y(t)$ (Fig. 22.2). For the x-direction, the momentum is

$$\frac{m_0 u_x}{\sqrt{1 - \left(\frac{u_x}{c}\right)^2 - \left(\frac{u_y}{c}\right)^2}} = \frac{m_0 c \beta_{u_x}}{\sqrt{1 - \beta_{u_x}^2 - \beta_{u_y}^2}} = qE_x t, \quad (22.11)$$

and for the y-direction

$$\frac{m_0 u_y}{\sqrt{1 - \left(\frac{u_x}{c}\right)^2 - \left(\frac{u_y}{c}\right)^2}} = \frac{m_0 c \beta_{u_y}}{\sqrt{1 - \beta_{u_x}^2 - \beta_{u_y}^2}} = p_0 = \text{const.}$$

$$(22.12)$$

We now have to solve these two equations with the two unknowns β_{u_x} and β_{u_y}. The formalism can be reduced a little bit by

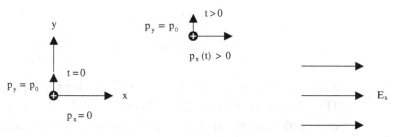

Fig. 22.2. A charge at the origin at $t = 0$ has a momentum p_0 in the y-direction. As of this moment, it is accelerated by an E-field in the x-direction.

introducing two abbreviations,

$$\frac{qE_x t}{m_0 c} = \Theta$$

$$\frac{p_0}{m_0 c} = \Pi.$$

Then we solve (22.11) for $\beta_{u_x}^2$,

$$\beta_{u_x}^2 = \Theta^2 \left(1 - \beta_{u_x}^2 - \beta_{u_y}^2 \right)$$

$$\beta_{u_x}^2 = \frac{\Theta^2 \left(1 - \beta_{u_y}^2 \right)}{1 + \Theta^2}, \tag{22.13}$$

and insert it into in (22.12),

$$\beta_{u_y}^2 = \Pi^2 \left[1 - \frac{\Theta^2 \left(1 - \beta_{u_y}^2 \right)}{1 + \Theta^2} - \beta_{u_y}^2 \right]$$

$$\beta_{u_y}^2 = \frac{\Pi^2 \left(1 - \dfrac{\Theta^2}{1 + \Theta^2} \right)}{1 - \Pi^2 \left(\dfrac{\Theta^2}{1 + \Theta^2} - 1 \right)} = \frac{\dfrac{\Pi^2}{1 + \Theta^2}}{1 + \dfrac{\Pi^2}{1 + \Theta^2}} = \frac{\Pi^2}{1 + \Pi^2 + \Theta^2}. \tag{22.14}$$

Then we insert (22.14) into (22.13),

$$\beta_{u_x}^2 = \frac{\Theta^2 \left(1 - \dfrac{\Pi^2}{1 + \Pi^2 + \Theta^2} \right)}{1 + \Theta^2} = \frac{\Theta^2}{1 + \Pi^2 + \Theta^2}. \tag{22.15}$$

Sketches of equations (22.14) and (22.15), for the special case of $\Pi = 1$, are shown in Fig. 22.3. As expected, β_{u_x} first increases linearly with time and then approaches unity asymptotically. Unexpected and utterly interesting, however, is the progression of β_{u_y}.

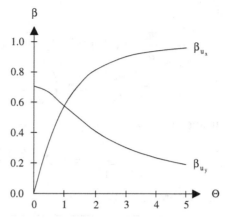

Fig. 22.3. Sketch of the functions (22.14) and (22.15) for $\Pi = 1$.

The charged particle's *velocity decreases* in a direction perpendicular to the electric field, a direction where *no force is acting!* The y-component of (22.6),

$$\gamma_u m_0 a_y = -\frac{f_x u_x}{c^2} u_y,$$

explains this phenomenon: Obviously, there is an acceleration a_y without a force f_y collinear with it. The momentum conservation law for the y-direction, (22.12), can also serve to explain the unexpected behavior of (22.14): As u_x increases, u_y has to decrease.

Let us now calculate the equation for the *particle's trajectory*. With $dt = \frac{m_0 c}{q E_x} d\Theta$, one writes for the velocity components

$$\beta_{u_x} = \frac{dx}{c\,dt} = \frac{q E_x}{m_0 c^2} \frac{dx}{d\Theta} \tag{22.16}$$

$$\beta_{u_y} = \frac{dy}{c\,dt} = \frac{q E_x}{m_0 c^2} \frac{dy}{d\Theta}, \tag{22.17}$$

equating (22.16) with the square root of (22.15), and (22.17) with the square root of (22.14), respectively, and integrating. The result is

$$x = \frac{m_0 c^2}{q E_x} \left(\sqrt{1 + \Pi^2 + \Theta^2} - \sqrt{1 + \Pi^2} \right) \qquad (22.18)$$

$$y = \frac{p_0 c}{q E_x} \operatorname{arsinh} \left(\frac{\Theta}{\sqrt{1 + \Pi^2}} \right). \qquad (22.19)$$

We rearrange (22.19),

$$\frac{\Theta}{\sqrt{1 + \Pi^2}} = \sinh \left(\frac{q E_x y}{p_0 c} \right),$$

and insert it into (22.18) to find

$$x = \frac{m_0 c^2}{q E_x} \left(\sqrt{(1 + \Pi^2) \left[1 + \sinh^2 \left(\frac{q E_x y}{p_0 c} \right) \right]} - \sqrt{1 + \Pi^2} \right)$$

$$x = \frac{m_0 c^2}{q E_x} \sqrt{1 + \Pi^2} \left[\cosh \left(\frac{q E_x y}{p_0 c} \right) - 1 \right]. \qquad (22.20)$$

The charge is moving on a *catenary*, a curve that has the shape of a chain suspended between two fixed points. For very small initial velocities, $u_y(0) \ll c$,

$$\Pi^2 = \left(\frac{p_0}{m_0 c} \right)^2 \approx \left(\frac{m_0 u_y(0)}{m_0 c} \right)^2 \to 0$$

holds, for very short times, $t = \frac{y}{u_y(0)} \ll \frac{m_0 c}{q E_x}$, the hyperbolic cosine is approximated by

$$\cosh \left(\frac{q E_x y}{m_0 u_y(0) c} \right) \approx 1 + \frac{1}{2} \left(\frac{q E_x y}{m_0 u_y(0) c} \right)^2.$$

This gives the non-relativistic result, the *trajectory parabola*,

$$x \approx \frac{q E_x}{2 m_0 u_y^2(0)} y^2. \qquad (22.21)$$

Chapter 23

Energy

When a force f_x is exerted on a mass point along a path element dx, the work done on it, $f_x\,dx$, gets stored as *kinetic energy* dE_{kin}. When the mass point has the rest mass m_0 and moves at a velocity of $\mathbf{u} = (u, 0, 0)$, we obtain with (22.5), i.e. the relation between force f_x and acceleration $a_x = du/dt$,

$$dE_{\text{kin}} = f_x\,dx = \gamma_u^3 m_0 \frac{du}{dt} dx = \gamma_u^3 m_0 \frac{dx}{dt} du = \frac{m_0 u\,du}{(1 - \beta_u^2)^{3/2}}$$

$$= m_0 c^2 \frac{\beta_u d\beta_u}{(1 - \beta_u^2)^{3/2}}.$$

If the motion starts with $u = 0$, integration yields

$$E_{\text{kin}} = m_0 c^2 \int_0^{\beta_u} \frac{\beta_u d\beta_u}{(1 - \beta_u^2)^{3/2}} = m_0 c^2 \left. \frac{1}{\sqrt{1 - \beta_u^2}} \right|_0^{\beta_u}$$

$$E_{\text{kin}} = \frac{m_0 c^2}{\sqrt{1 - \beta_u^2}} - m_0 c^2 = (\gamma_u - 1) m_0 c^2 = mc^2 - m_0 c^2. \quad (23.1)$$

This equation is essential! It says that, in relativistic physics, *kinetic energy is the difference of two energies*, namely, *the total energy, or relativistic energy,*

$$E = mc^2 = \frac{m_0 c^2}{\sqrt{1 - \left(\dfrac{u}{c}\right)^2}} = \frac{m_0 c^2}{\sqrt{1 - \beta_u^2}} = \gamma_u m_0 c^2, \quad (23.2)$$

and the *rest energy,* or *rest-mass energy,*

$$E_0 = m_0 c^2. \tag{23.3}$$

For non-relativistic velocities, β_u approaches zero, thus

$$\gamma_u = (1 - \beta_u^2)^{-1/2} \approx 1 + \frac{1}{2}\beta_u^2$$

$$E_{\text{kin}} \approx \left(1 + \frac{1}{2}\beta_u^2 - 1\right) m_0 c^2 = \frac{1}{2}m_0 u^2,$$

i.e. the well known non-relativistic result.

Equation (23.3) is probably the most famous equation of the theory of relativity: Any object of rest mass m_0 is endowed with a rest energy E_0; *rest mass and rest energy are equivalent.* Of particular interest is the huge proportionality coefficient, $c^2 = 9 \times 10^{16}\,\text{m}^2/\text{s}^2$, which assigns an inconceivably large amount of energy, $9 \times 10^{16}\,\text{J}$, to a mass as little as 1 kg. It would suffice to catapult a small mountain of mass 10^6t to the moon. Unfortunately – or perhaps fortunately – a sizeable fraction of the rest energy, in most cases all of it, cannot be converted to any other form of energy and vice versa. Exceptions may be found in high energy laboratories where, for example, a proton and an antiproton are formed *ex nihilo* when two protons collide at very high velocities. Conversely, a proton and an antiproton may *annihilate,* their masses convert to energy emitted as gamma radiation.

An important equation of mechanics can now be generalized with the help of (23.1). We start by differentiating with respect to time,

$$\frac{dE_{\text{kin}}}{dt} = \frac{dm}{dt}c^2, \tag{23.4}$$

write down the relativistic definition of force,

$$\mathbf{f} = \frac{d\mathbf{p}}{dt} = \frac{d}{dt}(m\mathbf{u}) = \frac{dm}{dt}\mathbf{u} + m\frac{d\mathbf{u}}{dt},$$

and insert (23.4),

$$\mathbf{f} = \frac{1}{c^2}\frac{dE_{\text{kin}}}{dt}\mathbf{u} + m\frac{d\mathbf{u}}{dt}.$$

Replacing $m = \gamma_u m_0$ and $\mathbf{a} = \frac{d\mathbf{u}}{dt}$ and rearranging terms yields

$$\gamma_u m_0 \mathbf{a} = \mathbf{f} - \frac{1}{c^2}\frac{dE_{\text{kin}}}{dt}\mathbf{u}. \tag{23.5}$$

Comparison of (22.6) and (23.5) finally gives

$$P = \frac{dE_{\text{kin}}}{dt} = \mathbf{f} \cdot \mathbf{u}. \tag{23.6}$$

The temporal change of the kinetic energy of a body, or the power P it absorbs, is the scalar product of force and velocity. This result is already known from classical mechanics – quite obviously relativity does not change it. We will need this result when we generalize force to four dimensions in a later chapter.

Total energy (or relativistic energy) not only comprises rest energy (or rest-mass energy) plus kinetic energy, but also *potential energy* E_{pot}. Let us think of a system consisting of two bodies, a and b, which interact electromagnetically. Their rest energies are $m_a c^2$ and $m_b c^2$. They shall move at the velocities u_a and u_b relative to their common center of mass. These motions raise their relativistic energies, when measured relative to the center of mass, to $m_a c^2/\sqrt{1 - (u_a/c)^2}$ and $m_b c^2/\sqrt{1 - (u_b/c)^2}$, respectively.[a] When potential energy, E_{pot}, is finally added which, as a matter of course, is negative for a bound system, the expression for the total energy reads

$$E = \frac{m_a c^2}{\sqrt{1 - \left(\dfrac{u_a}{c}\right)^2}} + \frac{m_b c^2}{\sqrt{1 - \left(\dfrac{u_b}{c}\right)^2}} + E_{\text{pot}}. \tag{23.7}$$

An example for such a two-body system is the *hydrogen atom* made of a proton p and an electron e. One decomposes the first two summands into two sums of the respective rest energies and kinetic energies

---

[a]It is interesting to note that *the heating of an object*, being nothing but an increase of the velocities of its constituent particles, *entails an increase of that body's mass*.

and writes

$$E = m_{\mathrm{p}}c^2 + E_{\mathrm{kin,p}} + m_{\mathrm{e}}c^2 + E_{\mathrm{kin,e}} + E_{\mathrm{pot}}.$$

Rearrangement gives

$$E = (m_{\mathrm{p}} + m_{\mathrm{e}})c^2 + (E_{\mathrm{kin,p}} + E_{\mathrm{kin,e}} + E_{\mathrm{pot}}). \qquad (23.8)$$

The second summand, consisting of both kinetic energies and the potential energy, is called *binding energy*. The ground state of atomic hydrogen has the well-known value of $-13.6\,\mathrm{eV}$. The proton's and the electron's rest energies, however, amount to $938\,\mathrm{MeV}$ and $511\,\mathrm{keV}$, respectively. In total, the hydrogen atom's total energy (and thus its total mass) is slightly smaller than the sum of the rest energies (rest masses) of its constituents. This effect is called *mass defect*. For the hydrogen isotope $_1^1\mathrm{H}$, it amounts to $13.6/(938 \times 10^6 + 511 \times 10^3) = 1.45 \times 10^{-6}\%$. This minute fraction of the total energies of proton and electron is released through emission of a photon when the hydrogen atom is formed by recombination of a proton and an electron.

When a nucleus is formed from protons and neutrons, electromagnetic interaction is accompanied by strong interaction. The mass defect is now much more evident than in the above example of atomic physics. Let us think of the $_2^4\mathrm{He}$ nucleus. It consists of two protons and two neutrons. One expects a mass of $2(m_{\mathrm{p}} + m_{\mathrm{n}}) = 2 \times (1.6726 + 1.6749) \times 10^{-27}\,\mathrm{kg} = 6.6950 \times 10^{-27}\,\mathrm{kg}$. However, its mass is as low as $6.6467 \times 10^{-27}\,\mathrm{kg}$. Therefore, the mass defect amounts to $(6.6950 - 6.6467)/6.6950 = 0.72\%$.

In general, *fusion* – i.e. the formation of light nuclei from protons, neutrons, and other light nuclei – is associated with energy release; the newly formed nucleus has a lower total energy (or total mass) than its constituents. *Exothermic nuclear reactions* of this kind can be found in many stars, the simplest one, turning hydrogen into helium, in hydrogen bombs as well. The composition of heavy nuclei through *endothermic nuclear reactions* occurs in nature only when supernovae explode. However, nuclei thus formed can afterwards be decomposed into lighter nuclei, a process also associated with energy release. It is called *fission*. The most famous example is the fission of uranium and plutonium in nuclear reactors and in nuclear bombs. Incidentally, the energy released in the Hiroshima bomb was equivalent to a mass defect of as little as 1 g!

Let us return once more to the problem calculated in Chapter 12, namely the *motion of a rocket travelling at a constant proper acceleration* α, i.e. constant in the frame of the accelerated observer. We can now find the increase of *kinetic energy of a mass m_0 as a function of the time elapsed since takeoff, the energy-time law*. Of course, kinetic energy E_{kin} and time t are measured in the S-frame which is at rest. Insertion of the velocity-time law, (12.7),

$$\frac{u_x}{c} = \frac{u}{c} = \beta_u = \frac{\dfrac{\alpha t}{c}}{\sqrt{1 + \left(\dfrac{\alpha t}{c}\right)^2}},$$

into (23.1),

$$E_{\text{kin}} = m_0 c^2 \left(\frac{1}{\sqrt{1 - \beta_u^2}} - 1 \right),$$

yields

$$E_{\text{kin}} = m_0 c^2 \left(\left(1 - \frac{\left(\dfrac{\alpha t}{c}\right)^2}{1 + \left(\dfrac{\alpha t}{c}\right)^2} \right)^{-1/2} - 1 \right)$$

$$= m_0 c^2 \left(\left(\frac{1}{1 + \left(\dfrac{\alpha t}{c}\right)^2} \right)^{-1/2} - 1 \right)$$

$$E_{\text{kin}} = m_0 c^2 \left(\sqrt{1 + \left(\frac{\alpha t}{c}\right)^2} - 1 \right). \tag{23.9}$$

For short times, $\alpha t \ll c$ holds, thus

$$E_{\text{kin}} \approx m_0 c^2 \left[1 + \frac{1}{2} \left(\frac{\alpha t}{c} \right)^2 - 1 \right] = \frac{1}{2} m_0 (\alpha t)^2 \approx \frac{1}{2} m_0 (a_x t)^2 \quad (23.10)$$

This is the non-relativistic result with E_{kin} growing proportionally to t^2. When E is plotted versus t in a double-logarithmic diagram, a straight line with a slope of two appears. For long times, $\alpha t \gg c$ holds, giving

$$E_{\text{kin}} \approx m_0 c^2 \sqrt{\left(\frac{\alpha t}{c} \right)^2} = m_0 c (\alpha t). \quad (23.11)$$

In this case, E_{kin} grows proportionally to t, a double-logarithmic diagram shows a slope of unity. With numerical values set to $\alpha = 10 \, \text{m/s}^2$ and $m_0 = 1 \, \text{kg}$, as in Chapter 12, we obtain Fig. 23.1 (note that the horizontal and vertical scales are different).

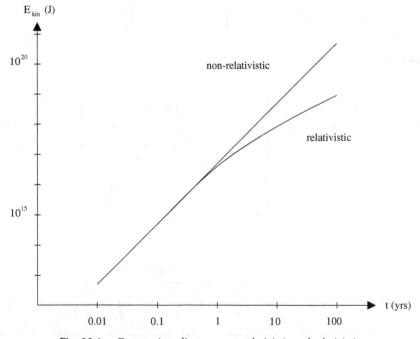

Fig. 23.1.   Energy-time diagram: non-relativistic and relativistic.

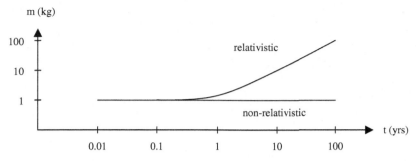

Fig. 23.2 Mass-time diagram: non relativistic and relativistic.

For the sake of completeness, we also want to derive the *dependence of relativistic mass on time, the mass-time law*. Inserting (12.7), the velocity-time law, into (20.1), the mass equation,

$$m = \frac{m_0}{\sqrt{1 - \beta_u^2}},$$

gives

$$m = m_0 \sqrt{1 + \left(\frac{\alpha t}{c}\right)^2}. \tag{23.12}$$

For long times,

$$m \approx m_0 \frac{\alpha t}{c},$$

so the mass grows linearly with time (see Fig. 23.2).

Chapter 24

Energy-Momentum Equation and Diagrams

In non-relativistic physics, the *relation between energy and momentum* of a mass point of mass m and velocity $\mathbf{u} = (u, 0, 0)$ is a relation between its kinetic energy E_{kin} and its momentum $\mathbf{p} = (p, 0, 0)$,

$$
E_{\text{kin}} = \frac{1}{2} m u^2 = \frac{p^2}{2m}. \tag{24.1}
$$

In relativistic physics, however, there is a corresponding relation between total energy E, rest energy E_0, and momentum p which can be derived as follows:

$$
E^2 = m^2 c^4 = \gamma_u^2 m_0^2 c^4 = m_0^2 c^4 \frac{1 - \beta_u^2 + \beta_u^2}{1 - \beta_u^2} = m_0^2 c^4 (1 + \gamma_u^2 \beta_u^2)
$$
$$
= m_0^2 c^4 + \gamma_u^2 m_0^2 \beta_u^2 c^4 = E_0^2 + m^2 u^2 c^2.
$$

With (20.2) we arrive at

$$
E^2 = E_0^2 + p^2 c^2 = (m_0 c^2)^2 + (pc)^2, \tag{24.2}
$$

or, as negative total energies do not occur,

$$
\frac{E}{c} = \sqrt{(m_0 c)^2 + p^2}. \tag{24.3}
$$

There are particles that are always moving at the speed of light, namely photons, and, at least to a very good approximation, *neutrinos.* Of course, this statement must be true for all inertial reference frames. But relativistic mass, given by (20.1), $m = m_0/\sqrt{1 - \beta_u^2}$, can be finite for $\beta_u = 1$ only when $m_0 = 0$. Therefore, the rest mass of particles moving at the speed of light is necessarily zero; they are called *massless*. For them the energy-momentum equation is reduced to

$$E = |p|\, c. \tag{24.4}$$

Of considerable practical importance is the *motion of an elementary particle* of rest mass m_0, charge q, and velocity u *perpendicular to a homogeneous magnetic field* of magnetic flux density, or magnetic induction, B. The Lorentz force, $f_L = quB$, makes the particle move on a circle. Its radius r can be calculated by equating the Lorentz force with the centripetal force. Non-relativistically, we write

$$quB = \frac{m_0 u^2}{r}. \tag{24.5}$$

The relativistic generalization is

$$quB = \frac{\gamma_u m_0 u^2}{r}, \tag{24.6}$$

as the rest mass m_0 has to be replaced by the relativistic mass $\gamma_u m_0$, according to (20.1). The Lorentz force expression, however, remains unchanged, as will be shown in Chapter 34. Using (20.2) to replace $\gamma_u m_0 u$ by the relativistic momentum p, (24.6) turns into

$$r = \frac{p}{qB}. \tag{24.7}$$

Since in most cases the total energy of a particle is given, we obtain with the help of (24.2),

$$r = \frac{\sqrt{E^2 - (m_0 c^2)^2}}{qBc}. \tag{24.8}$$

In *circular accelerators*, elementary particles can be accelerated to extremely relativistic energies, i.e. to $E \gg m_0 c^2$. This reduces (24.8) to

$$r \approx \frac{E}{qBc}. \tag{24.9}$$

For 300 GeV protons the ring requires a radius of 1 km and a magnetic induction of $B = 1$ T.

Conservation of energy and momentum are useful laws in elucidating the mechanics of point like particles considered in this text and, even more so, for systems of two or more interacting particles. The functional dependence of energy and momentum can be depicted in an *energy-momentum diagram*. In an *E/c-p*-diagram, particle *b* with a rest mass of $m_0 c^2$ is located on a *hyperbola branch* given by $E/c = \sqrt{(m_0 c)^2 + p^2}$. Massless particle *a* is found on either of two *straight lines* through the origin, $E/c = \pm p$, with slopes of ± 1 when the speed of light is set to $c = 1$. Since conserved quantities are plotted on both the vertical and the horizontal axis, arrows *a* and *b* can be *added like vectors* giving a new arrow, or vector, *s*, whose tip indicates energy and momentum of the two-body system (Fig. 24.1).

The laws of energy and momentum conservation allow all interactions in which the energy-momentum vectors of the two particles after interaction, *c* and *d*, add to the vector *s*. The only realization feasible is shown in Fig. 24.2. This becomes obvious when we try to move the tip of one vector

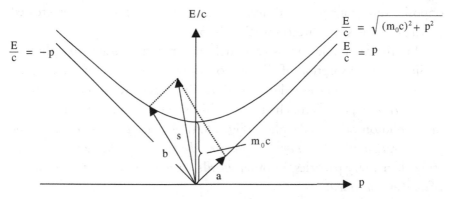

Fig. 24.1.   Energy-momentum diagram for two particles, *a* and *b*, without and with a rest mass, respectively.

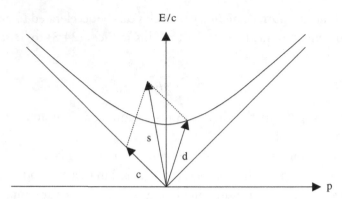

Fig. 24.2.   Energy-momentum diagram for particles *c* and *d*
after interaction of particles *a* and *b* of Fig. 24.1.

along the graph pertaining to it, only to find that *the other* vector's tip no
longer touches its particular graph. If, for example, we move the tip of *d*
along the hyperbola to the upper right, the tip of *c* leaves the bisector and
moves to the lower left. *c* would no longer be a photon vector. Figs. 24.1
and 24.2 depict the *one-dimensional Compton effect*: Photon *a* is moving
to the right and colliding with electron *b* approaching head-on (Fig. 24.1).
After the interaction, photon *c* is moving to the left and electron *d* to the
right (Fig. 24.2). Both particles have new energies and momenta.

Pair annihilation provides another example: Electron *a* and positron
b are converted into two photons, *c* and *d*. The two particles meet from
opposite directions and at different velocities, the photons thus created
depart into the same directions (Fig. 24.3).

In a third example, a massive particle at rest, *a*, tries to absorb a photon,
b. Since after absorption of the photon only one massive particle of the
same rest mass, *c*, is expected to be left, the tip of its vector should touch the
same *E/c-p*-hyperbola touched by *a*. This is not the case, thus absorption
of the photon cannot take place (Fig. 24.4). *Massive particles not allowing
the absorption of a photon are the electron, the proton, or the neutron*, i.e. all
truly elementary particles, in other words, *particles that are not composites
of smaller constituents*.

If the massive particle is composed of several smaller particles (e.g. all
atoms and all nuclei except $_1^1$H), *excited states* exist. These states possess

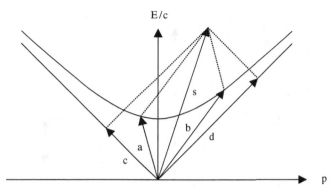

Fig. 24.3.   Energy-momentum diagram for pair annihilation:
particles *a* and *b* turn into photons *c* and *d*.

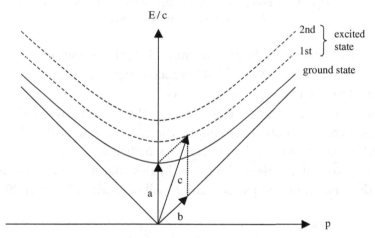

Fig. 24.4.   Energy-momentum diagram for the absorption
of a photon, *b*, by a particle at rest, *a*.

energies raised by the respective *excitation energies* and are thus described
by E/c-p-hyperbolae obtained by shifting the *ground state* hyperbola along
the E/c-axis. Now photons of suitable energies exist, capable of producing *c*-vectors whose tips touch one of the hyperbolae of an excited state.
Absorption can take place!

In a final example (Fig. 24.5), *a massive particle at rest, c, tries to emit
a photon, b*. Obviously, the massive particle must be in an excited state, it
must be a composite particle. Electrons, protons, and neutrons cannot emit

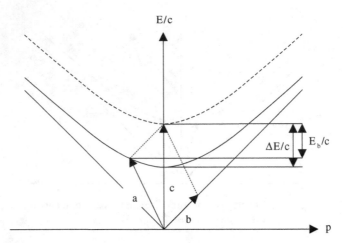

Fig. 24.5.   Energy-momentum diagram for the emission of a
photon, *b*, from a particle at rest, *c*.

photons. The energy of the photon emitted, E_b, is somewhat smaller than
the excitation energy, ΔE. The difference is equal to the kinetic energy
imparted to the massive particle as a recoil.

We would like to note at this point that the vertical scales of Figs. 24.4
and 24.5 are not linear: For example, the ground state of the H-atom is
938 MeV, but all of its excitation energies are below 13.6 eV.

Of course, all problems dealt with in this chapter can also be treated
quantitatively. Some sample calculation will be made in Chapter 29.

Chapter 25

Lorentz Transformation of Energy and Momentum

In this chapter we want to show that *energy and momentum behave in exactly the same way as time and space* when subjected to a Lorentz transformation, i.e. a transformation of certain sets of variables from one inertial reference frame to another one.

A mass point of rest mass m_0 is moving at $\mathbf{u} = (u, 0, 0)$ in the S-frame and at $\mathbf{u}' = (u', 0, 0)$ in the S'-frame, respectively. The relative velocity between the two inertial reference frames is $\mathbf{v} = (v, 0, 0)$. In the S-frame the total, or relativistic, energy is given by (23.2), and momentum by (20.2),

$$E = \frac{m_0 c^2}{\sqrt{1 - \left(\dfrac{u}{c}\right)^2}} = \gamma_u m_0 c^2 \qquad (25.1)$$

$$p_x = \gamma_u m_0 u = E \frac{u}{c^2}. \qquad (25.2)$$

For the calculations to follow, we need the γ-factor for u' as a function of u and v. It can be found with the addition theorem of velocities, (9.1),

$$u' = \frac{u - v}{1 - \dfrac{uv}{c^2}}$$

$$u'^2 = \frac{u^2 - 2uv + v^2}{\left(1 - \frac{uv}{c^2}\right)^2}$$

$$1 - \left(\frac{u'}{c}\right)^2 = \frac{1 - \frac{2uv}{c^2} + \frac{u^2v^2}{c^4} - \frac{u^2}{c^2} + \frac{2uv}{c^2} - \frac{v^2}{c^2}}{\left(1 - \frac{uv}{c^2}\right)^2}$$

$$= \frac{\left[1 - \left(\frac{u}{c}\right)^2\right]\left[1 - \left(\frac{v}{c}\right)^2\right]}{\left(1 - \frac{uv}{c^2}\right)^2}$$

$$\frac{1}{\sqrt{1 - \left(\frac{u'}{c}\right)^2}} = \frac{1 - \frac{uv}{c^2}}{\sqrt{1 - \left(\frac{u}{c}\right)^2}\sqrt{1 - \left(\frac{v}{c}\right)^2}}$$

$$\gamma_{u'} = \gamma_u \gamma_v \left(1 - \frac{uv}{c^2}\right). \tag{25.3}$$

We can now calculate energy and momentum in the S'-frame,

$$E' = \gamma_{u'} m_0 c^2 = \gamma_u \gamma_v \left(1 - \frac{uv}{c^2}\right) m_0 c^2 = \gamma_u \gamma_v \left(m_0 c^2 - m_0 uv\right)$$
$$= \gamma_v \left(E - vp_x\right),$$

$$p'_x = \gamma_{u'} m_0 u' = \gamma_u \gamma_v \left(1 - \frac{uv}{c^2}\right) m_0 \frac{u - v}{1 - \frac{uv}{c^2}} = \gamma_u \gamma_v \left(m_0 u - m_0 v\right)$$
$$= \gamma_v \left(p_x - \frac{v}{c^2} E\right).$$

By complete analogy to the Lorentz transformation for time and space, (6.5), we obtain the *Lorentz transformation for energy and momentum*:

$$\frac{E'}{c} = \gamma_v \left(\frac{E}{c} - \beta_v p_x \right) \qquad \frac{E}{c} = \gamma_v \left(\frac{E'}{c} + \beta_v p'_x \right)$$

$$p'_x = \gamma_v \left(p_x - \beta_v \frac{E}{c} \right) \qquad p_x = \gamma_v \left(p'_x + \beta_v \frac{E'}{c} \right)$$

$$p'_y = p_y \qquad\qquad\qquad p_y = p'_y$$

$$p'_z = p_z \qquad\qquad\qquad p_z = p'_z \qquad\qquad (25.4)$$

Energy, E/c, is Lorentz transformed exactly like time, ct, while the momentum components p_x, p_y, and p_z are transformed just like the spatial coordinates x, y, and z. Everything we have learned so far about Lorentz transformations and Minkowski diagrams in the *space-time domain* can now be applied one-to-one to the *energy-momentum domain*. Therefore, the Lorentz transformation in three spatial directions (18.6) can immediately be transcribed to

$$\frac{E'}{c} = \gamma_v \left(\frac{E}{c} - \boldsymbol{\beta}_v \cdot \mathbf{p} \right)$$

$$\mathbf{p}' = \mathbf{p} + (\gamma_v - 1) \frac{(\boldsymbol{\beta}_v \cdot \mathbf{p}) \boldsymbol{\beta}_v}{\beta_v^2} - \gamma_v \boldsymbol{\beta}_v \frac{E}{c}$$

$$\gamma_v = \frac{1}{\sqrt{1 - \beta_v^2}} = \frac{1}{\sqrt{1 - (\beta_x^2 + \beta_y^2 + \beta_z^2)}}. \qquad (25.5)$$

We can also depict the energy-momentum of an object a in a Minkowski diagram (Fig. 25.1) and answer the question of what its rest energy is in an inertial reference frame where it is at rest ($p'_x = 0$). To do so, we first calculate in the S-frame its distance from the origin as $\sqrt{\left(\frac{E}{c}\right)^2 + p_x^2}$ and remember that, following (7.2), a length unit in the S'-frame is $\sqrt{\frac{1+\beta_v^2}{1-\beta_v^2}}$.

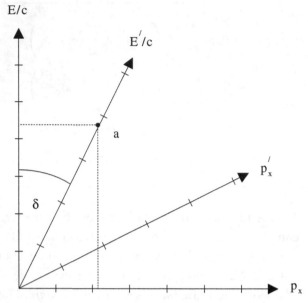

Fig. 25.1. Distance of an object *a* from the coordinate origin in an energy-momentum Minkowski diagram.

Thus, in the S'-frame, its distance from the coordinate origin is

$$\frac{E'}{c} = \frac{\sqrt{\left(\frac{E}{c}\right)^2 + p_x^2}}{\sqrt{\frac{1 + \beta_v^2}{1 - \beta_v^2}}}. \tag{25.6}$$

With (7.1) we write for the angle δ

$$\tan \delta = \frac{p_x}{\frac{E}{c}} = \beta_v, \tag{25.7}$$

insert this into (25.6) and obtain

$$\frac{E'}{c} = \frac{\sqrt{\left(\frac{E}{c}\right)^2 (1 + \beta_v^2)}}{\sqrt{\frac{1 + \beta_v^2}{1 - \beta_v^2}}} = \frac{1}{\gamma_v} \frac{E}{c}. \tag{25.8}$$

Of course, this result can immediately be written down with the first equation of the inverse Lorentz transformation (25.4, right-hand side) when all momentum components are set equal to zero. However, the most elegant solution for this problem will be undertaken in Chapter 28.

The Lorentz transformation of energy and momentum is also useful when calculating the center-of-mass velocity of a system of particles. Let us assume for simplicity that all particles move in the x-direction. Now all we have to do is postulate that the momenta of all particles in the center-of-mass frame, $p_x^{\text{c.o.m.}}$, add up to zero. With (25.4, left-hand side) we obtain

$$p_x^{\text{c.o.m.}} = 0 = \gamma_v \left(p_x - \beta_v^{\text{c.o.m.}} \frac{E}{c} \right)$$

$$\beta_v^{\text{c.o.m.}} = \frac{p_x c}{E}. \tag{25.9}$$

p_x and E are the sums of the energies and momenta, respectively, of all particles in the laboratory frame.

As an example of the above, let us look at a two-body system: a has a rest mass of m_a and the velocity $u_a \neq 0$, b has m_b and $u_b = 0$ in the laboratory frame. The momentum of the whole system amounts to

$$p_x = \frac{m_a u_a}{\sqrt{1 - \left(\dfrac{u_a}{c} \right)^2}},$$

the energy to

$$E = \frac{m_a c^2}{\sqrt{1 - \left(\dfrac{u_a}{c} \right)^2}} + m_b c^2.$$

This allows calculation of the center-of-mass velocity of the system,

$$v^{\text{c.o.m.}} = c \beta_v^{\text{c.o.m.}} = \frac{m_a u_a c^2}{m_a c^2 + m_b c^2 \sqrt{1 - \left(\dfrac{u_a}{c} \right)^2}} = \frac{u_a}{1 + \dfrac{m_b}{m_a} \sqrt{1 - \left(\dfrac{u_a}{c} \right)^2}}. \tag{25.10}$$

For the non-relativistic limit $u_a \to 0$, one obtains the known result,

$$v^{\text{c.o.m}} = \frac{m_a}{m_a + m_b} u_a. \tag{25.11}$$

With (25.10), energies and momenta of a and b in the center-of-mass frame are found by inserting $\beta_v^{c.o.m.}$ and $\gamma_v^{c.o.m.}$ into the Lorentz transformation.

With the help of the Lorentz transformation for energy and momentum even the *Planck relation*,

$$E = h\nu, \tag{25.12}$$

can be derived. It says that the energy E of a photon is proportional to its frequency ν and that the proportionality coefficient h is a universal constant.

Assume that in the S'-frame a photon of energy E' and momentum $(p_x', 0, 0) = (-E'/c, 0, 0)$ is being emitted along the negative x'-axis toward the origin of the S-frame. The inverse Lorentz transform gives

$$E = \gamma_v \left(E' + v p_x' \right) = \gamma_v \left(E' - \beta_v E' \right) = \frac{1 - \beta_v}{\sqrt{1 - \beta_v^2}} E',$$

or

$$E = \sqrt{\frac{1 - \beta_v}{1 + \beta_v}} E'. \tag{25.13}$$

In Chapter 14 (Doppler effect) we saw that a frequency ν' in the S'-frame is transformed to the frequency ν in the S-frame, as given by (14.3),

$$\nu = \sqrt{\frac{1 - \beta_v}{1 + \beta_v}} \nu'. \tag{25.14}$$

Division of (25.13) with (25.14) yields

$$\frac{E}{\nu} = \frac{E'}{\nu'}. \tag{25.15}$$

The energy of a photon divided by its frequency engenders the same numerical value in all inertial reference frames. Such quantities are said to be *Lorentz invariant*, and will be seen to be of great importance in Part IV, where the Lorentz invariance of certain variables and products will be dealt with in more detail. Experimentally, this proportionality coefficient, called *elementary quantum of action* or *Planck's constant*, h, is found to be

$$h = \frac{E}{\nu} = 6.63 \times 10^{-34} \, \text{Js}.$$

Chapter 26

Invariance and Distance in Three-Dimensional Space

A rod of length ℓ is located in a *two-dimensional space spanned by orthogonal coordinates*. The position of the origin of coordinates is arbitrary. Fig. 26.1 shows this rod in three such coordinate frames, $(x|y)$, $(x'|y')$, and $(x''|y'')$. All frames share the same origin, are not in motion relative to each other, but differ with respect to the direction of their axes. The frames are said to have been rotated around the origin.

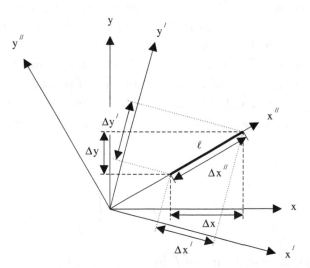

Fig. 26.1. A rod of length ℓ in three mutually rotated orthogonal coordinate frames.

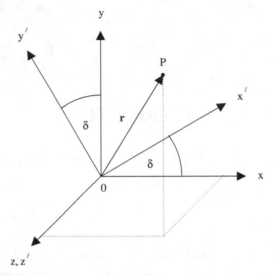

Fig. 26.2. Distance of a point P from the coordinate origin, 0.

With Pythagoras we write

$$\ell = \sqrt{\Delta x^2 + \Delta y^2} = \sqrt{\Delta x'^2 + \Delta y'^2} = \Delta x''. \qquad (26.1)$$

Obviously, *the length ℓ of the rod is the same in all mutually rotated coordinate frames*; it is called *rotationally invariant*. In the $(x''|y'')$-coordinate frame, the length is particularly easy to calculate as $\Delta y'' = 0$.

As a matter of fact, rotational invariance can be *generalized to three-dimensional space*. Let us take, e.g., the distance of point P from the coordinate origin (Fig. 26.2). This distance is the same as the length of the position vector $\mathbf{r}$ of point P. $\mathbf{r}$ is directed from 0 to P and is given in the $(x|y|z)$-frame by

$$\mathbf{r} = (x, y, z) \quad or \quad \mathbf{r} = \begin{pmatrix} x \\ y \\ z \end{pmatrix}. \qquad (26.2)$$

With (17.1), the coordinates of P in the $(x'|y'|z')$-frame, generated from the $(x|y|z)$-frame through rotation around the z-axis by an angle δ, are given by

$$x' = x \cos \delta + y \sin \delta$$
$$y' = -x \sin \delta + y \cos \delta$$
$$z' = z, \tag{26.3}$$

or in matrix notation,

$$\begin{pmatrix} x' \\ y' \\ z' \end{pmatrix} - \begin{pmatrix} \cos \delta & \sin \delta & 0 \\ -\sin \delta & \cos \delta & 0 \\ 0 & 0 & 1 \end{pmatrix} \begin{pmatrix} x \\ y \\ z \end{pmatrix}. \tag{26.4}$$

Let us now form the scalar product of the position vector with itself in the $(x|y|z)$-frame,

$$\mathbf{r} \cdot \mathbf{r} = (x, y, z) \cdot (x, y, z) = x^2 + y^2 + z^2. \tag{26.5}$$

This is the square of the distance from 0 to P. In the rotated $(x'|y'|z')$-frame, we obtain with (26.3),

$$\mathbf{r'} \cdot \mathbf{r'} = (x', y', z') \cdot (x', y', z') = x'^2 + y'^2 + z'^2$$
$$= x^2 \cos^2 \delta + 2xy \sin \delta \cos \delta + y^2 \sin^2 \delta$$
$$+ x^2 \sin^2 \delta - 2xy \sin \delta \cos \delta + y^2 \cos^2 \delta + z^2$$
$$= x^2 + y^2 + z^2$$
$$= \mathbf{r} \cdot \mathbf{r},$$

or

$$\mathbf{r} \cdot \mathbf{r} = \mathbf{r'} \cdot \mathbf{r'}. \tag{26.6}$$

The scalar product of the position vector with itself is rotationally invariant.

The momentum $\mathbf{p}$ of a body is given by a momentum vector which can be written by complete analogy to (26.2) as

$$\mathbf{p} = (p_x, p_y, p_z) \quad \text{or} \quad \mathbf{p} = \begin{pmatrix} p_x \\ p_y \\ p_z \end{pmatrix}. \tag{26.7}$$

If one rotates the coordinate frame around the z-axis, its momentum is written in new coordinates,

$$\mathbf{p}' = (p'_x, p'_y, p'_z) \quad \text{or} \quad \mathbf{p}' = \begin{pmatrix} p'_x \\ p'_y \\ p'_z \end{pmatrix}. \tag{26.8}$$

(26.8) is produced from (26.7) with the rotation matrix given in (26.4). After identical steps one finds that

$$\mathbf{p} \cdot \mathbf{p} = \mathbf{p}' \cdot \mathbf{p}'. \tag{26.9}$$

Therefore, *the scalar product of the momentum vector with itself is rotationally invariant*, as well. This result is already known, of course, as non-relativistic energy, i.e. kinetic energy, is given by

$$E_{\text{kin}} = \frac{p^2}{2m} = \frac{p_x^2 + p_y^2 + p_z^2}{2m} = \frac{\mathbf{p} \cdot \mathbf{p}}{2m}. \tag{26.10}$$

Being a scalar quantity, kinetic energy does not depend on the direction of the coordinate axes.

The catalog of rotationally invariant quantities can be continued. All scalar products must be rotationally invariant because scalar products are scalars, as the word says. To be sure, scalars are completely characterized by one single number that does not change upon rotation of the coordinate frame. The *vectors* making up the products, however, *are not rotationally invariant*, because they assume different positions in rotated coordinate frames and are thus composed of different sets of numbers.

Chapter 27

Invariance and Distance in Four-Dimensional Space-Time

Figure 27.1 is a Minkowski diagram for three inertial reference frames, S, S', and S'', spanned by the coordinate axes $(ct|x)$, $(ct'|x')$, and $(ct''|x'')$. All frames share the same origin. Two events, E_1 and E_2, connected by a line segment, are included. Such a line segment E_1E_2 is called *distance in four-dimensional space-time*.

The projections of E_1E_2 on the respective coordinate axes differ from reference frame to reference frame. The S-frame, however, is a special frame

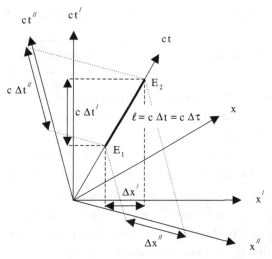

Fig. 27.1. Two events, E_1 and E_2, in three inertial reference frames, S, S', and S''.

for $E_1 E_2$ as the projection on the x-axis vanishes, $\Delta x = 0$. In this frame, E_1 and E_2 occur at the same point in space. The distance ℓ in space-time is just $\ell = c\Delta t = c\Delta\tau$; $\Delta\tau$ is the proper time between the events. As we already know, $c\Delta\tau$ is the shortest of all projections of the line segment $E_1 E_2$ on a time axis, i.e. $\Delta\tau$ is the shortest possible time interval between the events E_1 and E_2. With (2.4) we obtain

$$\Delta\tau = \Delta t'\sqrt{1 - \left(\frac{v'}{c}\right)^2} = \Delta t''\sqrt{1 - \left(\frac{v''}{c}\right)^2} = \dots \qquad (27.1)$$

$\Delta\tau$, $\Delta t'$, $\Delta t''$, ... are the time intervals in the S-, S'-, S''-...-frames, and v', v'', ... are the relative velocities between S and S', S and S'', ... We multiply (27.1) by c, put $\Delta t'$, $\Delta t''$, ... under the square roots and obtain the space-time distance ℓ,

$$\ell = c\Delta\tau = \sqrt{(c\Delta t')^2 - (v'\Delta t')^2} = \sqrt{(c\Delta t'')^2 - (v''\Delta t'')^2} = \dots$$

With $v' = \Delta x'/\Delta t'$, $v'' = \Delta x''/\Delta t''$, ... this expression is simplified to

$$\boxed{\ell^2 = c^2\Delta\tau^2 = c^2\Delta t'^2 - \Delta x'^2 = c^2\Delta t''^2 - \Delta x''^2 = \dots \qquad (27.2)}$$

(27.2) can be generalized to all four space-time dimensions,

$$\boxed{\ell^2 = c^2\Delta\tau^2 = c^2\Delta t'^2 - (\Delta x'^2 + \Delta y'^2 + \Delta z'^2) = \dots \qquad (27.3)}$$

With the exception of the minus sign, this equation is identical to Pythagoras's theorem. The *space-time distance* ℓ, calculated with (27.2) or (27.3), is *the same in all inertial reference frames*, while the lengths of the projections onto the respective space-time axes differ from reference frame to reference frame. Quantities that do not depend on the reference frame in four-dimensional space-time, thus being *invariant*, are of crucial importance in the chapters to follow. They are called *Lorentz invariant*.

According to (26.1), a length in three-dimensional space always has to be positive. With (27.2), however, *three possibilities* exist for a distance in space-time (Fig. 27.2):

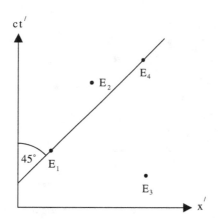

Fig. 27.2. Four events with time-like (E_1, E_2), space-like (E_1, E_3), and light-like (E_1, E_4) separations.

Case 1: $\ell^2 > 0 \rightarrow c|\Delta t'| > |\Delta x'|,\ c|\Delta t''| > |\Delta x''|, \ldots$

The *temporal distance* of two events, E_1 and E_2, is *larger than their spatial distance.* This is true for all inertial reference frames. For events E_1 and E_2, $|\Delta x'|/|\Delta t'|, \ldots < c$ is true; *these two events can happen to one and the same object*; therefore, the line segment $E_1 E_2$ can be the world line of an object. A *causal connection* of the events E_1 and E_2 can exist (meaning that E_1 can cause E_2). Two such events in space-time are said to have a *time-like separation.*

Case 2: $\ell^2 < 0 \rightarrow c|\Delta t'| < |\Delta x'|,\ c|\Delta t''| < |\Delta x''|, \ldots$

The *spatial distance* of two events, E_1 and E_3, is *larger than their temporal distance.* This is true for all inertial reference frames. For events E_1 and E_3, $|\Delta x'|/|\Delta t'|, \ldots > c$ is true; *these two events cannot happen to one object*, and $E_1 E_3$ cannot be a world line; *there cannot be a causal connection.* Two such events in space-time are said to have a *space-like separation.*

Case 3: $\ell^2 = 0 \rightarrow c|\Delta t'| = |\Delta x'|,\ c|\Delta t''| = |\Delta x''|, \ldots$

The *spatial distance and the temporal distance* of two events, E_1 and E_4, are *equal* – irrespective of the reference frame. For events E_1 and E_4, $|\Delta x'|/|\Delta t'|, \ldots = c$ is true. The line segment $E_1 E_4$ can be the *world line of light; a causal connection can thus exist.* Two such events are said to have a *light-like separation.*

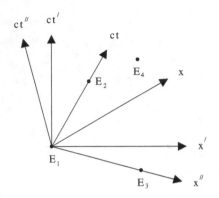

Fig. 27.3.   As in Fig. 27.2, but the origin is in E_1.

For the sake of simplicity, let us now shift the coordinate origin to E_1 (Fig. 27.3). We have already seen that for time-like distances, like E_1E_2, an x-ct-coordinate frame can be found that makes the spatial projection x vanish. In such a frame, both events, E_1 and E_2, are at the same locus. Such a frame is the rest frame for E_1E_2. The temporal projection ct is simply the proper time multiplied by c, i.e. $ct = c\tau$. Space-like distances, like E_1E_3, can be described by an x''-ct''-coordinate frame which makes the temporal projection ct'' vanish. In such a frame, events E_1 and E_3 are simultaneous, and the spatial projection x'' is nothing but the proper length of that distance.

Let us return for a while to Minkowski diagrams and ask *how a straight world line* of length ℓ from E_1 to E_2 *differs from any other world line* connecting these two events. Let us take, e.g., a transit traverse consisting of two line segments of lengths ℓ_1 and ℓ_2 from E_1 via E_3 to E_2 (Fig. 27.4).

In the S-frame, E_1 and E_2 have the same spatial coordinates, and so their distance in space-time is only a temporal one. Therefore, ℓ equals c multiplied by the proper time τ elapsing in S between the events E_1 and E_2. Correspondingly, the world lines from E_1 to E_3 and from E_3 to E_2 are equal to c multiplied by proper times τ_1 and τ_2 of coordinate frames (not shown in the figure) with time axes parallel to the line segments E_1E_3 and E_3E_2, respectively.

In the S-frame, the transit traverse $E_1E_3E_2$ describes a journey at very high velocities from a start to a finish and back. From the twin paradox

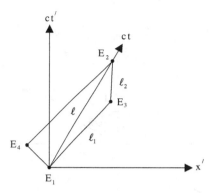

Fig. 27.4. Two transit traverses, $E_1 E_3 E_2$ and $E_1 E_4 E_2$.

(Chapter 15) we already know that the sum of two such proper times, $\tau_1 = \ell_1/c$ and $\tau_2 = \ell_2/c$, is smaller than the proper time of the straight connection, $\tau = \ell/c$. Of course, this is also true for all transit traverses consisting of more than two line segments.

The transit traverse $E_1 E_4 E_2$ describes the journey of a light ray from E_1 via a mirror at E_4 back to E_2. As explained in Case 3 (see above), the space-time distances for the two line segments are zero each, i.e. the proper time elapsed is zero.

All this leads us to an unexpected result: In four-dimensional space-time, represented in two dimensions by a Minkowski diagram, *a straight connection of two events, E_1 and E_2, with a time-like separation gives the longest, not the shortest, distance; the proper time elapsing is at a maximum. The shortest proper time, zero, elapses on a transit traverse* consisting of line segments aligned parallel to the two bisectors, in other words, *consisting of light-like world lines only*. Since a free particle, a particle not subject to external forces, always moves on a straight line between two events, such a particle ages most rapidly. This is called the *principle of maximum aging*.

Finally, we would like to mention that (27.2) is a very useful tool for the construction of Minkowski diagrams. We rewrite it in the form

$$\ell^2 = c^2 t'^2 - x'^2 = c^2 t^2 - x^2 = \ldots$$

by moving one end of the space-time distance ℓ into the coordinate origin and replacing double-primed variables by unprimed ones. We also set ℓ^2

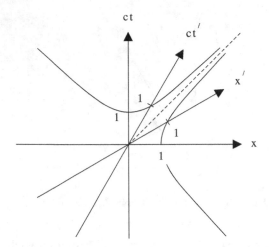

Fig. 27.5. Gauge hyperbolae: The intersection of the ct'-axis with the upper branch of the unit hyperbola gives the length unit on the ct'-axis. The length unit on the x'-axis is constructed in a similar way.

equal to unity and obtain

$$(ct)^2 - x^2 = (ct')^2 - x'^2 = 1. \tag{27.4}$$

$(ct)^2 - x^2 = 1$ is the equation of a unit hyperbola the upper branch of which (the one opened toward the top) has been drawn in Fig. 27.5.

The point $(ct' = 1 | x' = 0)$ satisfies (27.4); therefore it is a point of the unit hyperbola with unprimed variables. This point, however, is simply the unit vector in the ct'-direction of any S'-frame. Therefore, when the ct'-axis of any S'-frame is included in the diagram of the unit hyperbola, the intersection of that ct'-axis with the unit hyperbola automatically provides the unit length in the ct'-direction. With an analogous procedure the unit length on the x'-axis can be constructed. Since unit hyperbolae enable an immediate determination of the length units of any pair of coordinate axes, they are called *gauge hyperbolae*.

Problems for Part III

1. **Mass**

   What velocity is required for an object's mass to increase by a factor of 10; 100; 1000? Also, make a Taylor expansion for $m \to \infty$.

2. **Space travel**

   Subject is a one-way journey to a destination 180 ly away, calculated with (21.16). Make all calculations for $\tau = 0$ a; 1 a; 2 a; ... 10 a.

   (a) Take (12.13) to calculate the distance in the laboratory frame as a function of proper time.

   (b) Take (12.14) to calculate the velocity in the laboratory frame as a function of proper time.

   (c) Insert (12.12) into (12.9) and calculate the acceleration in the laboratory frame as a function of proper time.

   (d) Take (21.15) to calculate the mass of the spaceship in the reference frame of the spaceship as a function of proper time.

   (e) Calculate the spaceship's mass (for $\tau = 0$ a; 1 a; 2 a; ... 5 a) in the laboratory frame by inserting (21.15) and (12.14) into (20.1).

   (f) Insert (12.12) into (23.12) and calculate the mass of an astronaut in the laboratory frame as a function of proper time.

3. **Force**

   An object has a mass of 1.00 kg and is in motion at a velocity of $\mathbf{u} = (u, 0, 0)$, with $u = 0.01c; 0.10c; 0.50c; 0.90c; 0.99c$. Calculate its acceleration $\mathbf{a}$, when the force applied is (a) $\mathbf{f} = (1.00\,\text{N}, 0, 0)$, and (b) $\mathbf{f} = (0, 1.00\,\text{N}, 0)$, respectively.

4. **Energy I**

An electron at rest is very far from a proton ("at infinity"). The electron is attracted by the Coulomb force and approaches the proton in accelerated motion, until it is about 10^{-15} m from its center. The proton is much more massive than the electron, and the electron shall not emit radiation. Write down the non-relativistic and relativistic laws of energy conservation and solve them for the velocity. Calculate the velocity of the electron for 10^{-10} m $> r > 10^{-15}$ m.

5. **Energy II**

A circular disk of density ρ, thickness D, and radius R rotates around its axis at the angular velocity ω.

(a) Calculate its non-relativistic rotational energy (i.e. its kinetic energy) by integrating over circular rings from its center to its perimeter.

(b) Repeat the calculation in a relativistic way. What is the maximum for E_{kin}?

(c) Show with a Taylor expansion that the result of (a) can be derived from (b).

6. **Mass defect at nuclear fission**

The equation $^{235}_{92}\text{U} + ^{1}_{0}\text{n} \rightarrow ^{89}_{36}\text{Kr} + ^{144}_{56}\text{Ba} + 3^{1}_{0}\text{n}$ gives an example for nuclear fission. With the atomic mass unit $u = 1.6606 \times 10^{-27}$ kg, the masses are: uranium $- 235.04u$; krypton $- 88.91u$; barium $- 143.89u$; neutron $- 1.0087u$. Show that this nuclear reaction sets free an energy of about 200 MeV.

7. **Compton effect**

Photon a in Fig. 24.1 has an energy of $E_{\text{a}} = 237$ keV, electron b has $E_{\text{b}} = 632$ keV. Calculate the momenta pertaining to these energies, the vector $\mathbf{s} = (E/c, p)$, and thus the components of the vectors c and d in Fig. 24.2. For the sake of simplicity, use keV/c as the unit of momentum. Hint: You need 4 equations for the variables $E_{\text{c}}, E_{\text{d}}, p_{\text{c}}, p_{\text{d}}$; the electron's rest energy is 511 keV.

8. **Absorption of an X-ray photon by a nucleus**
   Nucleus *a* in Fig. 24.4 has an energy of $E_a = 60$ GeV, the gamma-ray photon *b* has $E_b = 1.00$ GeV. Calculate the excitation energy and the kinetic energy of the nucleus after absorption of the photon.

9. **Emission of a photon by an atom**
   The excited atom *c* in Fig. 24.5 has an energy of $E_c = 100$ GeV and emits a photon *b* with $E_b = 100$ keV. Calculate the kinetic energy and the velocity of the atom as a result of the recoil. Calculate both relativistically and non-relativistically.

10. **Lorentz transformation of energy and momentum**
    A particle of rest mass m_0 is moving in the *S*-frame at a velocity of $\mathbf{u} = (c/2, c/2, c/2)$.

    (a) Calculate its energy E and its momentum $\mathbf{p}$ in the *S*-frame.
    (b) Calculate E' and $\mathbf{p}'$ in the *S'*-frame moving at $\mathbf{v} = (c/2, 0, 0)$ relative to *S*.
    (c) Calculate E'' and $\mathbf{p}''$ in the *S''*-frame moving at $\mathbf{v} = (c/2, c/2, c/2)$ relative to *S*.

11. **Three-dimensional invariance**
    A particle of rest mass m_0 is moving in the *S*-frame at a velocity of $\mathbf{u} = (u_x, u_y, u_z)$. The *S'*-frame is obtained from rotation around the *z*-axis by the angle δ. Show that the particle's non-relativistic and relativistic kinetic energy is rotationally invariant, but that the particle's velocity is not.

12. **Four-dimensional invariance**
    Verify (27.2), i.e. the Lorentz invariance of the quantity $c^2 t'^2 - x'^2$, by inserting the Lorentz transformation (6.5). Analogously, verify $(E'/c)^2 - p_x'^2$.

PART IV
Four-Vectors and Electromagnetism

Chapter 28

Four-Vectors and Scalar Products

In preparation for the calculations of the following chapters, we have to build a formalism around equation (27.3) which, at first glance, seems redundant and even somewhat forbidding – the so-called *four-vector formalism*. It will be indispensable, however, in general relativity. For this reason, the reader should get used to it early.

Let us first define a so-called *contravariant four-dimensional time-position vector* by writing

$$X^\mu = (X^0, X^1, X^2, X^3) = (ct, x, y, z), \qquad (28.1)$$

and, correspondingly, a so-called *covariant four-dimensional time-position vector*,

$$X_\mu = (X_0, X_1, X_2, X_3) = (ct, -x, -y, -z). \qquad (28.2)$$

For four-vectors we use *capital letters*. While three-vectors are indicated by an arrow $(\vec{a})$ or by boldface (**a**), four-vectors appear *with a Greek superscript* (28.1) *or subscript* (28.2). The time component is called *zeroth component* ($\mu = 0$); the three spatial components are called *first, second, and third components* ($\mu = 1; 2; 3$), respectively. The only difference between contravariant and covariant vectors are the changes in sign of the first, second, and third components.

Let us now define the *scalar product of a contravariant with a covariant four-vector*, e.g. of the contravariant with the covariant time-position vector, by writing

$$X^\mu X_\mu = X^0 X_0 + X^1 X_1 + X^2 X_2 + X^3 X_3$$
$$= (ct, x, y, z)(ct, -x, -y, -z)$$
$$= (ct)^2 - x^2 - y^2 - z^2. \tag{28.3}$$

Note the *minus signs! No dot appears* between X^μ and X_μ, as with three-dimensional scalar products, or dot products, such as $\mathbf{r} \cdot \mathbf{r}$.

We remember the Lorentz transformation which can be written with a set of four equations, (6.5),

$$ct' = \gamma_v (ct - \beta_v x)$$
$$x' = \gamma_v (x - \beta_v ct)$$
$$y' = y$$
$$z' = z.$$

In accordance with (17.4), it can also be written with a four-by-four matrix,

$$\begin{pmatrix} ct' \\ x' \\ y' \\ z' \end{pmatrix} = \begin{pmatrix} \gamma_v & -\gamma_v \beta_v & 0 & 0 \\ -\gamma_v \beta_v & \gamma_v & 0 & 0 \\ 0 & 0 & 1 & 0 \\ 0 & 0 & 0 & 1 \end{pmatrix} \begin{pmatrix} ct \\ x \\ y \\ z \end{pmatrix}.$$

Let us now calculate the scalar product $X'^\mu X'_\mu$:

$$X'^\mu X'_\mu = (ct')^2 - x'^2 - y'^2 - z'^2$$
$$= \gamma_v^2 (ct - \beta_v x)^2 - \gamma_v^2 (x - \beta_v ct)^2 - y^2 - z^2$$
$$= \gamma_v^2 (ct)^2 (1 - \beta_v^2) - \gamma_v^2 x^2 (1 - \beta_v^2) - y^2 - z^2$$
$$= (ct)^2 - x^2 - y^2 - z^2$$
$$= X^\mu X_\mu.$$

We can therefore conclude that

$$X^\mu X_\mu = X'^\mu X'_\mu. \tag{28.4}$$

This is a very important result! The *four-dimensional scalar product* of the time-position vector with itself (more specifically: the four-dimensional scalar product of the contravariant time-position vector with its covariant counterpart) *is shape-invariant*, i.e. the mathematical form of the product is identical in the *S*-frame and in the *S'*-frame – except for the primes. There are *no additional terms* in the *S'*-frame that do not appear in the *S*-frame. If the scalar product of the time-position vector of an event with itself is evaluated, the *result is only a single number* – a scalar (for which reason the scalar product has received its name) – *irrespective of the inertial frame* used for its calculation. Thus equation (28.4) is the mathematical formulation of the circumstance that *a distance in four-dimensional space-time* (more precisely: its square) *is Lorentz invariant*, regardless of the inertial reference frame chosen.

Let us now turn to *energy and momentum*. As we have seen in Chapter 25, the energy coordinate, E/c, transforms like the time coordinate, ct, and the momentum coordinates, p_x, p_y, and p_z, transform like the spatial coordinates, x, y, and z (see (25.4)). So let us introduce contravariant and covariant energy-momentum vectors by

$$P^\mu = (P^0, P^1, P^2, P^3) = \left(\frac{E}{c}, p_x, p_y, p_z\right) \qquad (28.5a)$$

$$P_\mu = (P_0, P_1, P_2, P_3) = \left(\frac{E}{c}, -p_x, -p_y, -p_z\right). \qquad (28.5b)$$

An important hint regarding the notation must be given here: Only four-vectors and their four components require a distinction between superscripts and subscripts. Components of three-vectors always appear with subscripts, even if they are the components of a contravariant four-vector.

The scalar product of the energy-momentum vector with itself is given by analogy with (28.3),

$$P^\mu P_\mu = P^0 P_0 + P^1 P_1 + P^2 P_2 + P^3 P_3 = \left(\frac{E}{c}\right)^2 - p_x^2 - p_y^2 - p_z^2. \qquad (28.6)$$

Of course, by analogy with (28.4), it can also be shown that

$$P^\mu P_\mu = P'^\mu P'_\mu. \tag{28.7}$$

(28.6) and (28.7) can be used to derive the energy-momentum equation (24.2). For that purpose, we first write (28.6) in an inertial frame where momentum does not vanish,

$$P^\mu P_\mu = \left(\frac{E}{c}\right)^2 - p_x^2 - p_y^2 - p_z^2,$$

and then in an inertial frame where momentum vanishes,

$$P'^\mu P'_\mu = \left(\frac{E'}{c}\right)^2 = (m_0 c)^2.$$

Of course, the total energy E' in such a system is equal to the rest energy $m_0 c^2$. Equating both expressions according to (28.7), we obtain

$$E^2 - (p_x^2 + p_y^2 + p_z^2)c^2 = (m_0 c^2)^2$$
$$E^2 = (m_0 c^2)^2 + (pc)^2,$$

Q.E.D.

We can also derive (25.8) quite elegantly. The scalar product of the energy-momentum vector with itself of the particle, a in Fig. 25.1, amounts to $(E'/c)^2$ in the S'-frame, and $(E/c)^2 - p_x^2$ in the S-frame. Both expressions are equal according to (28.7), hence

$$\left(\frac{E'}{c}\right)^2 = \left(\frac{E}{c}\right)^2 - p_x^2.$$

With (20.2) and (23.2), i.e. $p_x = mv = \frac{E}{c^2}v = \frac{E}{c}\beta_v$, we obtain

$$\left(\frac{E'}{c}\right)^2 = \left(\frac{E}{c}\right)^2 (1 - \beta_v^2)$$

$$\frac{E'}{c} = \frac{1}{\gamma_v}\frac{E}{c},$$

Q.E.D.

Let us *generalize the situation a bit more*. We define the energy-momentum vectors of *two particles, a and b*,

$$P_a^\mu = \left(P_a^0, P_a^1, P_a^2, P_a^3\right) = \left(\frac{E_a}{c}, p_{ax}, p_{ay}, p_{az}\right),$$

$$P_b^\mu = \left(P_b^0, P_b^1, P_b^2, P_b^3\right) = \left(\frac{E_b}{c}, p_{bx}, p_{by}, p_{bz}\right).$$

Their scalar product is

$$P_a^\mu P_{b\mu} = P_a^0 P_{b0} + P_a^1 P_{b1} + P_a^2 P_{b2} + P_a^3 P_{b3}$$

$$= \frac{E_a E_b}{c^2} - p_{ax}p_{bx} - p_{ay}p_{by} - p_{az}p_{bz}. \tag{28.8}$$

As a matter of fact, this scalar product is also invariant. Application of the inverse Lorentz transformation (25.4, right-hand side) to the energy-momentum vectors of both particles yields

$$P_a^\mu P_{b\mu} = \frac{E_a E_b}{c^2} - p_{ax}p_{bx} - p_{ay}p_{by} - p_{az}p_{bz}$$

$$= \gamma_v^2 \left(\frac{E_a'}{c} + \beta_v p_{ax}'\right)\left(\frac{E_b'}{c} + \beta_v p_{bx}'\right)$$

$$- \gamma_v^2 \left(p_{ax}' + \beta_v \frac{E_a'}{c}\right)\left(p_{bx}' + \beta_v \frac{E_b'}{c}\right) - p_{ay}'p_{by}' - p_{az}'p_{bz}'$$

$$= \gamma_v^2 \left[\frac{E_a' E_b'}{c^2}(1 - \beta_v^2) + \frac{E_a'}{c}(\beta_v p_{bx}' - \beta_v p_{bx}')\right.$$

$$\left.+ \frac{E_b'}{c}(\beta_v p_{ax}' - \beta_v p_{ax}') + p_{ax}'p_{bx}'(\beta_v^2 - 1)\right] - p_{ay}'p_{by}' - p_{az}'p_{bz}'$$

$$= \frac{E_a' E_b'}{c^2} - p_{ax}'p_{bx}' - p_{ay}'p_{by}' - p_{az}'p_{bz}'$$

$$= P_a'^\mu P_{b\mu}',$$

thus

$$P_a^\mu P_{b\mu} = P_a'^\mu P_{b\mu}'. \tag{28.9}$$

Consequently, it is of no importance whether this scalar product is evaluated in the laboratory frame, in the center-of-mass frame, in the rest frame of particle *a* or of particle *b*, or in any other inertial reference frame. We shall exploit this invariance to address a variety of problems in the next chapter.

For the derivation of (28.4), (28.7), and (28.9), one-dimensional, i.e. *x*-directional, Lorentz transformations were used. Those scalar products also remain invariant when a three-dimensional Lorentz transformation, (18.8), is applied.

This chapter is to finish with an example for the scalar product of the time-position vector with itself. Figure 28.1 shows the *ct*- and *x*-axes of an *S*-frame and the *ct'*- and *x'*-axes of the *S'*-frame (with $\beta = 0.5$) pertaining to it. Three events are included:

$$E_a(ct = 2\,\text{ly} \mid x = 2\,\text{ly}), \; E_b(ct = 1\,\text{ly} \mid x = 5\,\text{ly}), \; E_c(ct = 5\,\text{ly} \mid x = 5\,\text{ly}).$$

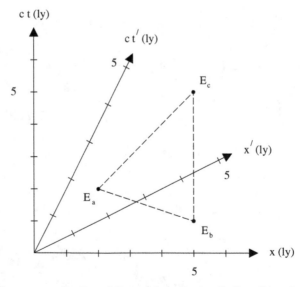

Fig. 28.1.   Three events, E_a, E_b and E_c, and their distances in four-dimensional space-time.

With the Lorentz transformation (6.5) one obtains the S'-coordinates of these three events

$$E_a\left(ct' = \frac{2}{\sqrt{3}}\,\mathrm{ly}\,\Big|\, x' = \frac{2}{\sqrt{3}}\,\mathrm{ly}\right), \quad E_b\left(ct' = -\sqrt{3}\,\mathrm{ly}\,\Big|\, x' = 3\sqrt{3}\,\mathrm{ly}\right),$$

$$E_c\left(ct' = \frac{5}{\sqrt{3}}\,\mathrm{ly}\,\Big|\, x' = \frac{5}{\sqrt{3}}\,\mathrm{ly}\right).$$

We would now like to see whether the distance of two events in four-dimensional space-time does not depend on the coordinate frames. To do so, we replace the absolute time-position vectors of (28.3) by difference vectors,

$$(X_a^\mu - X_b^\mu)(X_{a\mu} - X_{b\mu})$$

$$= (ct_a - ct_b, x_a - x_b, y_a - y_b, z_a - z_b)$$

$$\times (ct_a - ct_b, -(x_a - x_b), -(y_a - y_b), -(z_a - z_b))$$

or in short-form,

$$\Delta X^\mu \Delta X_\mu = (c\Delta t)^2 - \Delta x^2 - \Delta y^2 - \Delta z^2,$$

with $\Delta t = t_a - t_b$ and so on. For the pair of events E_a and E_b we obtain

$$\Delta X^\mu \Delta X_\mu = ((2-1)^2\,\mathrm{ly}^2 - (2-5)^2\,\mathrm{ly}^2 - 0^2\,\mathrm{ly}^2 - 0^2\,\mathrm{ly}^2) = -8\,\mathrm{ly}^2$$

$$\Delta X'^\mu \Delta X'_\mu$$

$$= \left(\left(\frac{2}{\sqrt{3}} + \sqrt{3}\right)^2\,\mathrm{ly}^2 - \left(\frac{2}{\sqrt{3}} - 3\sqrt{3}\right)^2\,\mathrm{ly}^2 - 0^2\,\mathrm{ly}^2 - 0^2\,\mathrm{ly}^2\right) = -8\,\mathrm{ly}^2.$$

The distance in space-time is the same in both frames. The minus sign indicates that the events have a space-like separation.

The reader is asked to show in a similar way that the distance squared between E_b and E_c amounts to $16\,\mathrm{ly}^2$ (time-like), and between E_a and E_c $0\,\mathrm{ly}^2$ (light-like).

Chapter 29

Calculus with the Energy-Momentum Vector

After having successfully completed the foregoing introduction into the ideas and the formalism of special relativity, we are now ready to tackle ten problems from modern physics that are less artificial than the bulk of the problems and thought experiments presented up to here.

Example 1: Completely Inelastic Collision of two Particles

Two particles, a and b, with rest masses m_a and m_b are moving in the laboratory frame with velocities u_a and u_b, respectively, in the x-direction. They collide completely inelastically to form a new particle, c. What is its rest mass, m_c, and its velocity, u_c?

Since we are looking for a relativistic solution, we write down the *energy conservation law for the total energy*,

$$E_a + E_b = E_c,$$

and also the *three momentum conservation laws*,

$$p_{ax} + p_{bx} = p_{cx}$$

$$p_{ay} + p_{by} = p_{cy}$$

$$p_{az} + p_{bz} = p_{cz},$$

and turn them into *a four-vector equation*,

$$\left(\frac{E_a}{c}, p_{ax}, p_{ay}, p_{az}\right) + \left(\frac{E_b}{c}, p_{bx}, p_{by}, p_{bz}\right) = \left(\frac{E_c}{c}, p_{cx}, p_{cy}, p_{cz}\right),$$

i.e.

$$P_a^\mu + P_b^\mu = P_c^\mu. \tag{29.1}$$

Constructing equations with *invariant scalar products* of the type (28.9) is our concern here. For that purpose, we multiply (29.1) by

$$P_{a\mu} + P_{b\mu} = P_{c\mu} \tag{29.2}$$

and obtain

$$P_a^\mu P_{a\mu} + P_b^\mu P_{b\mu} + P_a^\mu P_{b\mu} + P_b^\mu P_{a\mu} = P_c^\mu P_{c\mu}.$$

Since the third and fourth terms are equal (see (28.8)), the expression is simplified to

$$P_a^\mu P_{a\mu} + 2P_a^\mu P_{b\mu} + P_b^\mu P_{b\mu} = P_c^\mu P_{c\mu}. \tag{29.3}$$

We should now try to *find inertial reference frames where these scalar products assume the simplest forms possible* because, according to (28.9), we are free to choose suitable frames. In particular, we are completely free to choose different frames for each of the scalar products. This makes working with equations of the type (29.3) so advantageous.

For $P_a^\mu P_{a\mu}$, the rest frame of particle a is suitable because there the total energy, E_a, is just equal to the rest energy, $m_a c^2$, all momentum components vanish, and one obtains

$$P_a^\mu = \left(\frac{E_a}{c}, p_{ax}, p_{ay}, p_{az}\right) = (m_a c, 0, 0, 0).$$

Therefore,

$$P_a^\mu P_{a\mu} = (m_a c)^2. \tag{29.4}$$

$P_b^\mu P_{b\mu}$ and $P_c^\mu P_{c\mu}$ are calculated in the rest frames of particles b and c, respectively, yielding

$$P_b^\mu P_{b\mu} = (m_b c)^2 \tag{29.5}$$

$$P_c^\mu P_{c\mu} = (m_c c)^2. \tag{29.6}$$

The term $2P_a^\mu P_{b\mu}$ is calculated most favorably in the laboratory frame where the total energy of particle a amounts to

$$E_a = \gamma_a m_a c^2 = \frac{m_a c^2}{\sqrt{1 - \left(\frac{u_a}{c}\right)^2}},$$

a corresponding result being true for E_b. Hence we can write

$$P_a^\mu = (\gamma_a m_a c, \gamma_a m_a u_a, 0, 0)$$
$$P_b^\mu = (\gamma_b m_b c, \gamma_b m_b u_b, 0, 0),$$

and

$$2P_a^\mu P_{b\mu} = 2(\gamma_a m_a c, \gamma_a m_a u_a, 0, 0)(\gamma_b m_b c, -\gamma_b m_b u_b, 0, 0)$$
$$2P_a^\mu P_{b\mu} = 2\gamma_a \gamma_b m_a m_b (c^2 - u_a u_b). \tag{29.7}$$

(29.4), (29.7), (29.5), and (29.6), inserted into (29.3), yield

$$(m_a c)^2 + 2\gamma_a \gamma_b m_a m_b \left(c^2 - u_a u_b\right) + (m_b c)^2 = (m_c c)^2, \tag{29.8}$$

and, solved for the rest mass of the new particle c,

$$m_c = \sqrt{m_a^2 + m_b^2 + 2m_a m_b \gamma_a \gamma_b \left(1 - \frac{u_a u_b}{c^2}\right)}. \tag{29.9}$$

Let us now examine (29.9), more specifically the product of three factors in the third term under the square root,

$$\gamma_a \gamma_b \left(1 - \frac{u_a u_b}{c^2}\right) = \left[1 - \left(\frac{u_a}{c}\right)^2\right]^{-1/2} \left[1 - \left(\frac{u_b}{c}\right)^2\right]^{-1/2} \left(1 - \frac{u_a u_b}{c^2}\right).$$

For the non-relativistic limit, i.e. $u_a, u_b \ll c$, the product equals unity, thus $m_c = m_a + m_b$. For $u_a = 0$ and arbitrary u_b, the first and third factors are unity, the second factor is greater than unity. For $u_b = 0$ and arbitrary u_a, corresponding results hold. If u_a and u_b are arbitrary and have opposite signs, all three factors are greater than unity. If they carry equal signs (this case being possible as the velocities are evaluated in the laboratory frame!), it can be shown in a few lines that the product of all three factors is also greater than unity.

Consequently, in the relativistic limit the third term under the square root is always greater than $2m_a m_b$, therefore the *rest mass, m_c, of the new particle is always greater than the sum of the rest masses, m_a and m_b, of the two particles from which it was formed.* Their *kinetic energies* in the rest frame of the new particle have turned into *rest energy*, i.e. rest mass times c^2. If the new particle, c, were to decay again into particles a and b – as is the case with the fission of a heavy nucleus – a part of the rest energy of c would reappear as kinetic energies of a and b.

> There is no rest mass conservation law in relativistic physics.

The velocity u_c of the new particle c, in the laboratory frame can be calculated with the energy conservation law,

$$\gamma_a m_a c^2 + \gamma_b m_b c^2 = \gamma_c m_c c^2, \qquad (29.10)$$

and the momentum conservation law for the x-direction,

$$\gamma_a m_a u_a + \gamma_b m_b u_b = \gamma_c m_c u_c. \qquad (29.11)$$

When (29.10) is divided by c^2 and inserted into (29.11), one obtains

$$u_c = \frac{\gamma_a m_a u_a + \gamma_b m_b u_b}{\gamma_a m_a + \gamma_b m_b}. \qquad (29.12)$$

u_c is, of course, the center-of-mass velocity of the laboratory frame. Setting $u_b = 0$ and $\gamma_b = 1$, (25.10) is re-derived.

Example 2: Production of a Proton-Antiproton Pair

A very energetic proton, a, is being shot at proton b at rest. If its kinetic energy is sufficient, a proton-antiproton pair is produced. In symbols:

$$p + p \rightarrow p + p + (p + \bar{p}).$$

p and $\bar{p}$ denote proton and antiproton, respectively. When the energy applied is minimized, the four particles are at rest relative to each other after the collision, forming a cluster c. What minimum energy, E_a, and what velocity, u_a, are required for proton a?

Obviously, the situation of example 1 is almost completely replicated, the only differences being

$$u_b = 0 \rightarrow \gamma_b = 1$$
$$m_a = m_b = m_0$$
$$m_c = 4m_0,$$

where m_0 is the rest mass of the proton and antiproton. From (29.8) follows

$$(m_0 c)^2 + 2\gamma_a (m_0 c)^2 + (m_0 c)^2 = \left(4 m_0 c\right)^2$$
$$2\gamma_a m_0^2 c^2 = 14 m_0^2 c^2$$
$$E_a = 7 m_0 c^2. \tag{29.13}$$

A total energy equivalent of seven proton rest energies is needed for the proton in motion. The velocity u_a needed is

$$E_a = \frac{m_0 c^2}{\sqrt{1 - \left(\dfrac{u_a}{c}\right)^2}} = 7 m_0 c^2$$

$$\left(\frac{1}{7}\right)^2 = 1 - \left(\frac{u_a}{c}\right)^2$$

$$u_a = c\sqrt{1 - \left(\tfrac{1}{7}\right)^2} \approx 0.990c, \tag{29.14}$$

i.e. a velocity only 1.0% below the speed of light.

Energetically more favorable is a *storage ring* where *two protons with equal but opposite velocities* collide in order to create a proton-antiproton pair. When energies are minimized, the cluster consisting of three protons and one antiproton is again at rest in the laboratory frame after the collision. What are the protons' minimum energies, $E = E_a = E_b$, and velocities, $u_a = -u_b$, under these circumstances?

Let us again return to the setup of example 1 and replace the variables of (29.8) in the following manner:

$$u_a = u; \quad u_b = -u$$

$$\gamma_a = \gamma_b = \gamma_u$$

$$m_a = m_b = m_0$$

$$m_c = 4m_0.$$

Thence we obtain

$$(m_0 c)^2 + 2\gamma_u^2 m_0^2 (c^2 + u^2) + (m_0 c)^2 = (4m_0 c)^2$$

$$\gamma_u^2 (c^2 + u^2) = 7c^2$$

$$c^2 + u^2 = 7(c^2 - u^2)$$

$$u = \sqrt{\tfrac{3}{4}} c \approx 0.866 c. \tag{29.15}$$

A smaller velocity is required. Before the collision, each proton has an energy of

$$E = \frac{m_0 c^2}{\sqrt{1 - \tfrac{3}{4}}} = 2m_0 c^2, \tag{29.16}$$

i.e. a kinetic energy of $m_0 c^2$. So the kinetic energy of both protons totals $2m_0 c^2$. In the original experiment, the kinetic energy needed was $6m_0 c^2$; 2/3 of the energy is saved.

Example 3: Absorption of a Photon by an Atom (see also Fig. 24.4)

An atom a at rest with a rest mass of m_a absorbs a photon b propagating in the x-direction with a frequency of $v = E_b/h$, thus turning atom a into atom c. What is the atom's rest mass, m_c, and its velocity, u_c?

We start with (29.1), more specifically with its squared form, (29.3),

$$P_a^\mu P_{a\mu} + 2P_a^\mu P_{b\mu} + P_b^\mu P_{b\mu} = P_c^\mu P_{c\mu}.$$

For the rest frame of the atom before the absorption we write

$$P_a^\mu = (m_a c, 0, 0, 0)$$

$$P_b^\mu = \left(\frac{E_b}{c}, p_{bx}, 0, 0\right) = \left(\frac{E_b}{c}, \frac{E_b}{c}, 0, 0\right) = \left(\frac{hv}{c}, \frac{hv}{c}, 0, 0\right).$$

Here, the energy-momentum equation for photons (24.4) and the Planck relation (25.12) are used. For the rest frame of the atom after absorption one writes

$$P_c^\mu = (m_c c, 0, 0, 0).$$

The new aspect of this example is *the scalar product of the energy-momentum vector of a photon with itself.* It is

$$P_b^\mu P_{b\mu} = \left(\frac{hv}{c}\right)^2 - \left(\frac{hv}{c}\right)^2 = 0. \qquad (29.17)$$

(29.17) is *true in any inertial reference frame!* Inserting this into (29.3) gives

$$(m_a c)^2 + 2m_a c \frac{hv}{c} + 0 = (m_c c)^2 \qquad (29.18)$$

$$m_c = \sqrt{m_a^2 + 2\frac{m_a hv}{c^2}} = m_a \sqrt{1 + 2\frac{hv}{m_a c^2}}$$

$$\approx m_a \left[1 + \frac{hv}{m_a c^2} - \frac{1}{2}\left(\frac{hv}{m_a c^2}\right)^2\right]$$

$$m_c \approx m_a + \frac{hv}{c^2} - \frac{1}{2}m_a \left(\frac{hv}{m_a c^2}\right)^2. \qquad (29.19)$$

To first-order approximation, the *atom's rest mass increases* by the photon's mass $h\nu/c^2$. An increase of the rest mass is possible because the atom is a composite particle, composed of a nucleus and electron(s). The increase of the rest mass is manifested by a *decrease of the electron's binding energy*. Since binding energies are quantized, only photons with well-defined energies can be absorbed.

To second-order approximation, the mass increase is smaller by $\frac{1}{2}m_a\left(\frac{h\nu}{m_a c^2}\right)^2$ because the atom is set in motion. Equating this mass decrease with the non-relativistic kinetic energy, divided by c^2, we obtain

$$\frac{1}{2}m_a\left(\frac{h\nu}{m_a c^2}\right)^2 = \frac{1}{2}m_a\left(\frac{u_c}{c}\right)^2$$

$$u_c = \frac{h\nu}{m_a c}. \tag{29.20}$$

In the laboratory frame, the atom's exact velocity after the absorption is calculated with the momentum conservation law for the x-direction and the energy conservation law,

$$\frac{h\nu}{c} = \gamma_c m_c u_c \tag{29.21}$$

$$m_a c^2 + h\nu = \gamma_c m_c c^2. \tag{29.22}$$

Solving for $\gamma_c m_c$ and equating both left-hand sides yields

$$\frac{h\nu}{u_c c} = m_a + \frac{h\nu}{c^2}$$

$$u_c = \frac{h\nu c}{m_a c^2 + h\nu}. \tag{29.23}$$

To zeroth-order approximation in $\frac{h\nu}{m_a c^2}$, the result (29.20) is re-established:

$$u_c = \frac{h\nu c}{m_a c^2 \left(1 + \dfrac{h\nu}{m_a c^2}\right)} \approx \frac{h\nu}{m_a c}. \tag{29.24}$$

Example 4: Emission of a Photon by an Atom (see also Fig. 24.5)

Atom c at rest with a rest mass of m_c is emitting a photon b of frequency $v = E_b/h$ in the x-direction. After emission, the atom is called a and has a rest mass of m_a. What is the photon's energy, E_b, as a function of the difference of the rest energies, $\Delta E = E_c - E_a$?

The starting equation is again (29.1) which, when multiplied by $P_{a\mu}$, provides a new conservation law,

$$P_a^\mu P_{a\mu} + P_b^\mu P_{a\mu} = P_c^\mu P_{a\mu}. \tag{29.25}$$

Four energy-momentum vectors are needed:

- P_a^μ in the rest frame of a : $P_a^\mu = (m_a c, 0, 0, 0)$
- P_a^μ in the rest frame of c : $P_a^\mu = \left(\gamma_a m_a c, -\dfrac{E_b}{c}, 0, 0\right)$
- P_b^μ in the rest frame of c : $P_b^\mu = \left(\dfrac{E_b}{c}, \dfrac{E_b}{c}, 0, 0\right)$
- P_c^μ in the rest frame of c : $P_c^\mu = (m_c c, 0, 0, 0)$.

We calculate the first term of (29.25) in the rest frame of a, the two other terms in the rest frame of c, and obtain

$$m_a^2 c^2 + \gamma_a m_a E_b + \left(\frac{E_b}{c}\right)^2 = \gamma_a m_a m_c c^2. \tag{29.26}$$

The energy conservation law in the rest frame of c is

$$\gamma_a m_a c^2 + E_b = m_c c^2. \tag{29.27}$$

Insertion of (29.27) into (29.26) yields

$$m_a^2 c^2 + \frac{E_b}{c^2}(m_c c^2 - E_b) + \left(\frac{E_b}{c}\right)^2 = m_c^2 c^2 - m_c E_b$$

$$m_a^2 c^2 + m_c E_b = m_c^2 c^2 - m_c E_b$$

$$E_b = \frac{m_c^2 c^2 - m_a^2 c^2}{2m_c} = \frac{E_c^2 - E_a^2}{2E_c} = (E_c - E_a)\left(1 - \frac{E_c - E_a}{2E_c}\right).$$

With $\Delta E = E_c - E_a$, the final result is

$$E_b = \Delta E \left(1 - \frac{\Delta E}{2E_c}\right). \tag{29.28}$$

So the photon's energy is *not* equal to the difference of the atom's rest energies but is *less* because in the emission process the *atom experiences recoil, thus gaining kinetic energy at the expense of the photon.* The change, $\Delta E/2E_c$, decreases with increasing mass of the atom.

There is a good chance for the photon not to be absorbed by an atom of the same kind; resonance absorption becomes a problem. The conditions for *recoilless resonance absorption* to occur nonetheless, e.g. the re-absorption of a gamma-ray photon by a nucleus of the same kind, were investigated by Mössbauer. The *Mössbauer effect* has not only been of importance to experimentally verifying the general theory of relativity but is still frequently being used in materials research.[a]

Example 5: Decay of a Particle at Rest into two Particles

A particle c at rest with a rest mass of m_c decays into two particles, a and b, with rest masses m_a and m_b, respectively. What are the respective kinetic energies, $E_{a,kin}$ and $E_{b,kin}$, of the two particles generated?

The decay is the reverse of the completely inelastic collision (example 1). Therefore, (29.3) and (29.8) are valid again:

$$(m_a c)^2 + 2\gamma_a \gamma_b m_a m_b (c^2 - u_a u_b) + (m_b c)^2 = (m_c c)^2.$$

u_a and u_b are the velocities of a and b, respectively, in the laboratory frame, i.e. in the rest frame of c. In order to eliminate the mixed terms $u_a u_b$ and $\gamma_a \gamma_b$ with a second equation, we take (29.25),

$$P_a^\mu P_{a\mu} + P_b^\mu P_{a\mu} = P_c^\mu P_{a\mu}.$$

We choose the rest frame of a for the first term and obtain $(m_a c)^2$ again. For the two other terms we write down P_a^μ, P_b^μ, and P_c^μ in the laboratory

---

[a]In 1958 Rudolf L. Mössbauer first reported on recoilless gamma emission. He was awarded the Nobel Prize in 1961.

frame, requiring that a and b move in the x-direction,

$$P_a^\mu = (\gamma_a m_a c, \gamma_a m_a u_a, 0, 0)$$
$$P_b^\mu = (\gamma_b m_b c, \gamma_b m_b u_b, 0, 0)$$
$$P_c^\mu = (m_c c, 0, 0, 0).$$

This turns (29.25) into

$$(m_a c)^2 + \gamma_a \gamma_b m_a m_b (c^2 - u_a u_b) = \gamma_a m_a m_c c^2. \tag{29.29}$$

We subtract (29.29) twice from (29.8) to obtain

$$-(m_a c)^2 + (m_b c)^2 = (m_c c)^2 - 2\gamma_a m_a m_c c^2$$
$$2E_a m_c = (m_a^2 + m_c^2 - m_b^2)c^2$$
$$E_a = \frac{(m_a^2 + m_c^2 - m_b^2)c^2}{2m_c}$$
$$E_{a,kin} = \frac{(m_a^2 + m_c^2 - m_b^2)c^2}{2m_c} - m_a c^2$$
$$E_{a,kin} = \frac{[(m_c - m_a)^2 - m_b^2]c^2}{2m_c}. \tag{29.30a}$$

Interchanging the indices a and b gives

$$E_{b,kin} = \frac{[(m_c - m_b)^2 - m_a^2]c^2}{2m_c}. \tag{29.30b}$$

When (29.30a) and (29.30b) are added, we obtain

$$E_{kin} = E_{a,kin} + E_{b,kin} = (m_c - m_a - m_b)c^2. \tag{29.31}$$

So *the kinetic energy is equivalent to the mass defect* occurring with the decay of the particle.

Example 6: Production of an Electron-Positron Pair from a Photon (Pair Production)

A photon, a, is flying along the x-axis and decaying into an electron, c, and a positron, d. Both particles have the same rest mass, m_0. *What minimum energy, E_a, is required for the photon in the laboratory frame?*

This reaction is written symbolically as $h\nu \to e^- + e^+$. The energy-momentum conservation law is

$$P_a^\mu = P_c^\mu + P_d^\mu, \tag{29.32}$$

or when squared,

$$P_a^\mu P_{a\mu} = P_c^\mu P_{c\mu} + 2P_c^\mu P_{d\mu} + P_d^\mu P_{d\mu}. \tag{29.33}$$

According to (29.17), $P_a^\mu P_{a\mu} = 0$ holds. The first and third terms on the right-hand side are $(m_0 c)^2$ each. The second term is calculated in the center-of-mass frame, giving

$$P_c^\mu = \left(\frac{E_c}{c}, p_{cx}, p_{cy}, p_{cz} \right)$$

$$P_d^\mu = \left(\frac{E_d}{c}, -p_{dx}, -p_{dy}, -p_{dz} \right).$$

Of course, both particles have the same total energies, $E_c = E_d$, and the same magnitudes of momentum for all three directions. Therefore,

$$2P_c^\mu P_{d\mu} = 2 \left(\frac{E_c^2}{c^2} + p_{cx}^2 + p_{cy}^2 + p_{cz}^2 \right), \tag{29.34}$$

and (29.33) becomes

$$0 = 2(m_0 c)^2 + 2 \left(\frac{E_c^2}{c^2} + p_c^2 \right). \tag{29.35}$$

Obviously, this equation is *unsolvable. The photon cannot decay into a pair, at least not in a vacuum. It requires the presence of another particle* which shall have a rest mass of M and the index b before and e after the collision:

$$P_a^\mu + P_b^\mu = P_c^\mu + P_d^\mu + P_e^\mu. \qquad (29.36)$$

Solving this equation is somewhat tedious, so we restrict our discussion to the special case of three particles, c, d, and e, staying together and forming a cluster c,

$$P_a^\mu + P_b^\mu = P_c^\mu. \qquad (29.37)$$

The terms on the left-hand side are calculated in the laboratory frame where b is at rest,

$$P_a^\mu = \left(\frac{E_a}{c}, \frac{E_a}{c}, 0, 0 \right)$$

$$P_b^\mu = (Mc, 0, 0, 0).$$

P_c^μ is written in the rest frame of the cluster,

$$P_c^\mu = ((M + 2m_0)\, c, 0, 0, 0).$$

Thus (29.37), after squaring, is

$$0 + 2E_a M + (Mc)^2 = (M + 2m_0)^2 c^2$$

$$2E_a M = 4Mm_0 c^2 + 4m_0^2 c^2$$

$$E_a = 2m_0 c^2 + \frac{2m_0^2 c^2}{M}$$

$$E_a = 2m_0 c^2 \left(1 + \frac{m_0}{M} \right). \qquad (29.38)$$

Again, the absence of an extra mass M ($M \to 0$) would be unphysical because an undefined expression would result. The larger M is, the smaller is the additional energy required for pair production. If M is the mass of an electron, $M = m_0$, the photon's energy must be twice the size of the rest energy of the pair.

Example 7: Dematerialization of an Electron-Positron Pair into two Photons (Pair Annihilation) (see also Fig. 24.3)

Electron a is moving along the x-axis and has a total energy of E_a in the laboratory frame. It hits positron b, at rest in the laboratory frame (but in motion in Fig. 24.3). Both electron and positron have the same rest mass, m_0. They *decay into two photons*, c and d, which move in the x-direction. What are their energies, E_c and E_d?

We start again with the energy-momentum equation for the system consisting of four particles and square it,

$$P_a^\mu + P_b^\mu = P_c^\mu + P_d^\mu, \tag{29.39}$$

$$P_a^\mu P_{a\mu} + 2P_a^\mu P_{b\mu} + P_b^\mu P_{b\mu} = P_c^\mu P_{c\mu} + 2P_c^\mu P_{d\mu} + P_d^\mu P_{d\mu}. \tag{29.40}$$

The first and third terms on the left-hand side amount to $(m_0 c)^2$ each. The second term is calculated in the rest frame of the positron where

$$P_a^\mu = \left(\frac{E_a}{c}, p_{ax}, 0, 0 \right)$$

$$P_b^\mu = (m_0 c, 0, 0, 0)$$

holds. So the left-hand side reads $2(m_0 c)^2 + 2E_a m_0$.

The first and third terms on the right-hand side give zero each. The second term is calculated in the laboratory frame, yielding

$$P_c^\mu = \left(\frac{E_c}{c}, p_{cx}, 0, 0 \right)$$

$$P_d^\mu = \left(\frac{E_d}{c}, p_{dx}, 0, 0 \right).$$

So we find for the right-hand side

$$2 \left(\frac{E_c E_d}{c^2} - p_{cx} p_{dx} \right),$$

and (29.40) turns into

$$(m_0 c)^2 + E_a m_0 = \frac{E_c E_d}{c^2} - p_{cx} p_{dx}.$$

With $p_{cx} = \pm E_c/c$ and $p_{dx} = \pm E_d/c$, the right-hand side is non-zero only if p_{cx} and p_{dx} have different signs. Thus we obtain

$$(m_0 c)^2 + E_a m_0 = \frac{2}{c^2} E_c E_d. \qquad (29.41)$$

This gives us a first equation for E_c and E_d. A second one is obtained by multiplying (29.39) by $P_{b\mu}$,

$$P_a^\mu P_{b\mu} + P_b^\mu P_{b\mu} = P_c^\mu P_{b\mu} + P_d^\mu P_{b\mu}, \qquad (29.42)$$

and calculating all four products in the laboratory frame,

$$E_a m_0 + m_0^2 c^2 = E_c m_0 + E_d m_0$$
$$E_a + m_0 c^2 = E_c + E_d. \qquad (29.43)$$

This is simply energy conservation! We solve (29.43) for E_d and insert it into (29.41),

$$(m_0 c)^2 + E_a m_0 = \frac{2}{c^2} E_c (E_a + m_0 c^2 - E_c)$$

$$\frac{1}{2} m_0 c^2 (m_0 c^2 + E_a) = -E_c^2 + E_c (E_a + m_0 c^2).$$

This quadratic equation has two solutions,

$$E_{c_{1,2}} = \frac{1}{2}(E_a + m_0 c^2) \pm \sqrt{\frac{1}{4}(E_a + m_0 c^2)^2 - \frac{1}{2} m_0 c^2 (m_0 c^2 + E_a)}$$

$$= \frac{1}{2}(E_a + m_0 c^2) \pm \frac{1}{2}\sqrt{(E_a + m_0 c^2)(E_a - m_0 c^2)}.$$

When (29.43) is solved for E_c and inserted into (29.41), $E_{d_{1,2}} = E_{c_{1,2}}$ follows, thus

$$E_{c,d} = \frac{1}{2}(E_a + m_0 c^2) \pm \frac{1}{2}\sqrt{(E_a + m_0 c^2)(E_a - m_0 c^2)}. \qquad (29.44)$$

Example 8: Compton Effect (see also Fig. 24.1 and Fig. 24.2)

A photon of frequency v (before collision: index a) collides with an electron at rest of mass m_0 (before collision: index b; in Fig. 24.1 in motion, however). After the collision, the photon (index c) has released part of its energy. Now it has a frequency of v', leaving its original flight path at the scattering angle θ. The electron (index d, after collision) is also in motion (Fig. 29.1). What is the functional relation between v (or wavelength λ) and v' (or λ') with θ as a parameter?

We start once more with (29.39), or its squared form, (29.40). The first term and the fourth term (photon terms) are zero; the third term and the sixth term (electron terms) are equal in the respective rest frames of the electron. What remains is

$$P_a^\mu P_{b\mu} = P_c^\mu P_{d\mu}. \tag{29.45}$$

Multiplication of (29.39) by $P_{c\mu}$ gives

$$P_a^\mu P_{c\mu} + P_b^\mu P_{c\mu} = P_c^\mu P_{c\mu} + P_d^\mu P_{c\mu}.$$

The photon term disappears again, so

$$P_a^\mu P_{c\mu} + P_b^\mu P_{c\mu} = P_d^\mu P_{c\mu} \tag{29.46}$$

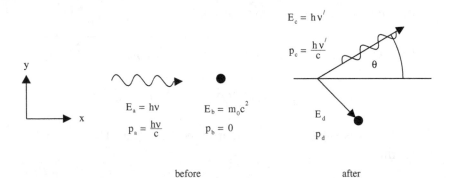

Fig. 29.1. Compton effect.

remains. From (29.45) and (29.46), $P_{d\mu}$ can be eliminated,

$$P_a^\mu P_{c\mu} + P_b^\mu P_{c\mu} = P_a^\mu P_{b\mu}. \tag{29.47}$$

In the laboratory frame

$$P_a^\mu = \left(\frac{h\nu}{c}, \frac{h\nu}{c}, 0, 0\right)$$

$$P_b^\mu = (m_0 c, 0, 0, 0)$$

$$P_c^\mu = \left(\frac{h\nu'}{c}, \frac{h\nu'}{c}\cos\theta, \frac{h\nu'}{c}\sin\theta, 0\right)$$

holds. This turns (29.47) into

$$\frac{h^2}{c^2}\nu\nu'(1 - \cos\theta) + m_0 h\nu' = m_0 h\nu. \tag{29.48}$$

With the fundamental wave-equation, $\nu = c/\lambda$, some more conversions are possible, yielding

$$\frac{h^2}{\lambda\lambda'}(1 - \cos\theta) + \frac{m_0 hc}{\lambda'} = \frac{m_0 hc}{\lambda}$$

$$\frac{h^2}{\lambda\lambda'}(1 - \cos\theta) = m_0 hc\frac{\lambda' - \lambda}{\lambda\lambda'}$$

$$\lambda' - \lambda = \frac{h}{m_0 c}(1 - \cos\theta). \tag{29.49}$$

This is the famous *Compton equation*. $h/m_0 c = 2.42 \times 10^{-12}$ m is known as the *Compton wavelength of the electron*.

Example 9: Inverse Compton Effect

The situation is the same as in example 8, except that the electron arrives at a very high velocity, $-u_b$, from the right. For simplicity, the photon is reflected back into the x-direction. We start again with (29.47). In the laboratory frame, we have

$$P_{\text{a}}^{\mu} = \left(\frac{h\nu}{c}, \frac{h\nu}{c}, 0, 0\right)$$

$$P_{\text{b}}^{\mu} = (\gamma_{\text{b}} m_0 c, -\gamma_{\text{b}} m_0 u_{\text{b}}, 0, 0)$$

$$P_{\text{c}}^{\mu} = \left(\frac{h\nu'}{c}, -\frac{h\nu'}{c}, 0, 0\right).$$

For the photon energies, we write $E = h\nu$ and $E' = h\nu'$, and obtain from (29.47)

$$\frac{2EE'}{c^2} + \gamma_{\text{b}} m_0 E' - \gamma_{\text{b}} m_0 \frac{u_{\text{b}}}{c} E' = \gamma_{\text{b}} m_0 E + \gamma_{\text{b}} m_0 \frac{u_{\text{b}}}{c} E. \qquad (29.50)$$

Since $u_{\text{b}} \approx c$ holds, we make the approximations $u_{\text{b}}/c \approx 1$, thus

$$1 - \frac{u_{\text{b}}}{c} \approx \frac{1}{2}\left(1 + \frac{u_{\text{b}}}{c}\right)\left(1 - \frac{u_{\text{b}}}{c}\right) = \frac{1}{2}\left(1 - \frac{u_{\text{b}}^2}{c^2}\right) = \frac{1}{2\gamma_{\text{b}}^2},$$

and obtain

$$\frac{2EE'}{c^2} + \frac{m_0 E'}{2\gamma_{\text{b}}} = 2\gamma_{\text{b}} m_0 E$$

$$\frac{E'}{E} = \frac{2\gamma_{\text{b}} m_0}{\dfrac{2E}{c^2} + \dfrac{m_0}{2\gamma_{\text{b}}}} = \frac{4\gamma_{\text{b}}^2}{1 + \dfrac{4\gamma_{\text{b}} E}{m_0 c^2}}. \qquad (29.51)$$

The inverse Compton effect is of importance in astronomy. Photons of the microwave background radiation, with a very low energy of the order of $E \approx 10^{-3}$ eV, are struck by very energetic electrons ($m_0 c^2 = 511$ keV). For typical values of $\gamma_{\text{b}} \gg 10^8$, the second term in the denominator dominates, giving

$$E' \approx \gamma_{\text{b}} \times 511 \text{ keV}. \qquad (29.52)$$

The inverse Compton effect can thus increase the energy of a photon in a spectacular manner.

Example 10: Bremsstrahlung

Electron a, having a rest mass of m_a and arriving from the left at a velocity of u_a, hits nucleus b at rest with a rest mass of m_b. After the collision, both particles form a cluster, c, which moves to the right. In this process, photon d is emitted to the left. What energy E does it have? As in some of the previous examples, the starting equation for this calculation is (29.39), or (29.40),

$$P_a^\mu P_{a\mu} + 2P_a^\mu P_{b\mu} + P_b^\mu P_{b\mu} = P_c^\mu P_{c\mu} + 2P_c^\mu P_{d\mu} + P_d^\mu P_{d\mu}.$$

The fifth term contains the velocity of the cluster. It has not been given nor is it asked for. This term is therefore to be eliminated. For this purpose, we multiply (29.39) by $2P_{d\mu}$ and obtain

$$2P_a^\mu P_{d\mu} + 2P_b^\mu P_{d\mu} = 2P_c^\mu P_{d\mu} + 2P_d^\mu P_{d\mu}. \tag{29.53}$$

Subtraction of (29.53) from (29.40) gives

$$P_a^\mu P_{a\mu} + 2P_a^\mu P_{b\mu} + P_b^\mu P_{b\mu} - 2P_a^\mu P_{d\mu} - 2P_b^\mu P_{d\mu} = P_c^\mu P_{c\mu} - P_d^\mu P_{d\mu}. \tag{29.54}$$

This equation has seven terms. The seventh is the photon term and thus zero. The first, third, and sixth terms are calculated in their respective rest frames, yielding $(m_a c)^2$, $(m_b c)^2$, and $(m_a + m_b)^2 c^2$. The second, fourth, and fifth terms are calculated in the laboratory frame, yielding

$$P_a^\mu = (\gamma_a m_a c, \gamma_a m_a u_a, 0, 0)$$

$$P_b^\mu = (m_b c, 0, 0, 0)$$

$$P_d^\mu = \left(\frac{E}{c}, -\frac{E}{c}, 0, 0\right).$$

This turns (29.54) into

$$(m_a c)^2 + 2\gamma_a m_a m_b c^2 + (m_b c)^2 - 2\gamma_a m_a (1 + \beta_a) E - 2m_b E$$
$$= (m_a + m_b)^2 c^2 + 0$$
$$2E[\gamma_a m_a (1 + \beta_a) + m_b] = 2(\gamma_a - 1) m_a m_b c^2$$

$$E = \frac{(\gamma_a - 1) m_a m_b c^2}{\gamma_a m_a (1 + \beta_a) + m_b}. \tag{29.55}$$

In clusters of galaxies, the intergalactic gas has temperatures of about 10^7 to 10^8 K. The kinetic energy of the particles (protons and electrons) is thus found to be between 1.3 and 13 keV. With $E_{kin} = (\gamma_a - 1) m_a c^2$, one obtains for electrons $\gamma_a = 1.0025 - 1.025$. So even for the lightest nuclei (protons) the first term in the denominator is negligible, and with $m_a c^2 = 511$ keV we get

$$E \approx (\gamma_a - 1) \times 511 \, \text{keV}. \tag{29.56}$$

Bremsstrahlung is thus X-radiation with energies of 1.3 to 13 keV, the kinetic energy of the electron is completely being converted into radiation.

Chapter 30*

Four-Velocity and Four-Acceleration

After the introduction of a four-dimensional time-position vector X^μ, construction of a *four-dimensional velocity vector* U^μ by taking the first derivative of X^μ with respect to t suggests itself. Such a choice would be unwise, however, as t is frame-dependent, i.e. not Lorentz invariant, and scalar products calculated this way would no longer be shape-invariant. For this reason, *differentiation with respect to the Lorentz invariant proper time τ is a better choice,*

$$U^\mu = \frac{dX^\mu}{d\tau}. \tag{30.1}$$

Four-velocity, denoted by upper-case letters, can be expressed as a function of the *real* velocity, denoted by lower-case letters,

$$U^\mu = (U^0, U^1, U^2, U^3) = \frac{dX^\mu}{dt}\frac{dt}{d\tau} = \gamma_u(c, u_x, u_y, u_z). \tag{30.2}$$

$dt = \gamma_u\, d\tau$ is the relation between the proper time τ of a process and the dilated time t, measured in any inertial reference frame where this process is in motion at a velocity of u. γ_u is given by (22.4).

At this point we would like to warn readers not to interpret U^μ as a measurable quantity, since dX^μ is the infinitesimal space-time distance between two events in one frame (e.g. the laboratory frame), while $d\tau$ is the infinitesimal time increment in a specific, totally different frame in which both events take place at the same location. The three spatial components $\gamma_u u_x$, $\gamma_u u_y$, and $\gamma_u u_z$ may even become arbitrarily large, while the time component $\gamma_u c$ is of little use to describe the events. What then is this four-vector good for? In what follows, it will only serve us as an aid to abbreviate and increase the elegance of calculations already done and also to derive new useful quantities.

Let us now *subject the four-velocity to a Lorentz transformation*. The problem to be solved is the following: In the S-frame, an object has a velocity of $\mathbf{u} = (u_x, u_y, u_z)$, and a four-velocity of $U^\mu = \gamma_u(c, u_x, u_y, u_z)$, with γ_u being $[1 - \frac{1}{c^2}(u_x^2 + u_y^2 + u_z^2)]^{-1/2}$. What is the velocity $\mathbf{u}' = (u_x', u_y', u_z')$, and the four-velocity $U'^\mu = \gamma_{u'}(c, u_x', u_y', u_z')$, with $\gamma_{u'}$ being $[1 - \frac{1}{c^2}(u_x'^2 + u_y'^2 + u_z'^2)]^{-1/2}$ in the S'-frame moving at the velocity $\mathbf{v} = (v, 0, 0)$ relative to the S-frame? Let us define $\gamma_v = (1 - \beta_v^2)^{-1/2}$ and use the concise matrix notation,

$$
U'^\mu = \begin{pmatrix} \gamma_{u'} c \\ \gamma_{u'} u_x' \\ \gamma_{u'} u_y' \\ \gamma_{u'} u_z' \end{pmatrix} = \begin{pmatrix} \gamma_v & -\gamma_v \beta_v & 0 & 0 \\ -\gamma_v \beta_v & \gamma_v & 0 & 0 \\ 0 & 0 & 1 & 0 \\ 0 & 0 & 0 & 1 \end{pmatrix} \begin{pmatrix} \gamma_u c \\ \gamma_u u_x \\ \gamma_u u_y \\ \gamma_u u_z \end{pmatrix}. \tag{30.3}
$$

We multiply (30.3) to obtain

$$
\gamma_{u'} c = \gamma_v \gamma_u c - \frac{v}{c} \gamma_v \gamma_u u_x \tag{30.4a}
$$

$$
\gamma_{u'} u_x' = -\frac{v}{c} \gamma_v \gamma_u c + \gamma_v \gamma_u u_x \tag{30.4b}
$$

$$
\gamma_{u'} u_y' = \gamma_u u_y \tag{30.4c}
$$

$$
\gamma_{u'} u_z' = \gamma_u u_z. \tag{30.4d}
$$

From (30.4a), an equation results which, if $u_y = u_z = 0$ holds, is identical with (25.3),

$$
\gamma_{u'} = \gamma_u \gamma_v \left(1 - \frac{u_x v}{c^2}\right). \tag{30.5}
$$

Inserted into (30.4b), one finds

$$
\left(1 - \frac{u_x v}{c^2}\right) u_x' = u_x - v,
$$

$$u'_x = \frac{u_x - v}{1 - \dfrac{u_x v}{c^2}}.$$

Correspondingly, we find for (30.4c) and (30.4d)

$$u'_y = \frac{u_y}{\gamma_v \left(1 - \dfrac{u_x v}{c^2}\right)},$$

$$u'_z = \frac{u_z}{\gamma_v \left(1 - \dfrac{u_x v}{c^2}\right)}.$$

The last three equations are (with $\gamma_v = \gamma$) identical with equations (9.1), the addition theorem of velocities. Obviously, the derivation of this important set of equations has become much faster and more transparent in this chapter.

To gain more confidence in this somewhat strange concept of a four-velocity, let us calculate the scalar product of four-velocity with itself,

$$U^\mu U_\mu = \gamma_u^2 (c, u_x, u_y, u_z)(c, -u_x, -u_y, -u_z) = \gamma_u^2 (c^2 - u_x^2 - u_y^2 - u_z^2)$$

$$= \frac{c^2 - u_x^2 - u_y^2 - u_z^2}{1 - \dfrac{1}{c^2}(u_x^2 + u_y^2 + u_z^2)} = c^2,$$

$$U^\mu U_\mu = c^2. \tag{30.6}$$

$U^\mu U_\mu$ is, as expected, a Lorentz invariant quantity, thus

$$U^\mu U_\mu = U'^\mu U'_\mu. \tag{30.7}$$

$U^\mu U_\mu$ may be calculated in any inertial reference frame, e.g. the one where $u_x = u_y = u_z = 0$ and $\gamma_u = 1$ is true, which yields directly c^2.

The following problem gives us an example of how to apply (30.7): In the S-frame, particles a and b have the velocities $\mathbf{u}_a$ and $\mathbf{u}_b$, respectively. What is the magnitude of the relative velocity $\mathbf{u}'_b$ of particle b in the

S'-frame, the rest frame of particle a? We write down the scalar product of both four-velocities in the S-frame,

$$U_a^\mu U_{b\mu} = U_a^0 U_{b0} - (U_a^1 U_{b1} + U_a^2 U_{b2} + U_a^3 U_{b3})$$

$$= \gamma_a \gamma_b [c^2 - (u_{ax} u_{bx} + u_{ay} u_{by} + u_{az} u_{bz})]$$

$$= \gamma_a \gamma_b (c^2 - \mathbf{u}_a \cdot \mathbf{u}_b). \qquad (30.8)$$

In the S'-frame, with $u'_{ax} = u'_{ay} = u'_{az} = 0$ and $\gamma'_a = 1$, we find

$$U_a'^\mu U_{b\mu}' = \gamma_b' c^2. \qquad (30.9)$$

Equating (30.8) and (30.9), followed by a division by c^2, leads to

$$\gamma_b' = \gamma_a \gamma_b \left(1 - \frac{\mathbf{u}_a \cdot \mathbf{u}_b}{c^2}\right).$$

The reciprocal squared is

$$1 - \left(\frac{u_b'}{c}\right)^2 = \frac{\left[1 - \left(\frac{u_a}{c}\right)^2\right]\left[1 - \left(\frac{u_b}{c}\right)^2\right]}{\left(1 - \frac{\mathbf{u}_a \cdot \mathbf{u}_b}{c^2}\right)^2}$$

$$u_b'^2 = c^2 \frac{\left(1 - \frac{\mathbf{u}_a \cdot \mathbf{u}_b}{c^2}\right)^2 - \left[1 - \left(\frac{u_a}{c}\right)^2\right]\left[1 - \left(\frac{u_b}{c}\right)^2\right]}{\left(1 - \frac{\mathbf{u}_a \cdot \mathbf{u}_b}{c^2}\right)^2}$$

$$= \frac{(\mathbf{u}_a - \mathbf{u}_b)^2 - \frac{u_a^2 u_b^2}{c^2} + \frac{(\mathbf{u}_a \cdot \mathbf{u}_b)^2}{c^2}}{\left(1 - \frac{\mathbf{u}_a \cdot \mathbf{u}_b}{c^2}\right)^2}$$

$$u_b'^2 = \frac{(\mathbf{u}_a - \mathbf{u}_b)^2 - \frac{(\mathbf{u}_a \times \mathbf{u}_b)^2}{c^2}}{\left(1 - \frac{\mathbf{u}_a \cdot \mathbf{u}_b}{c^2}\right)^2}. \qquad (30.10)$$

The identity $(\mathbf{a} \times \mathbf{b})^2 = a^2 b^2 - (\mathbf{a} \cdot \mathbf{b})^2$ was used in the last step. One can immediately see that, non-relativistically, (30.10) simplifies to $u_b' = |\mathbf{u}_a - \mathbf{u}_b|$.

Correspondingly, a *four-acceleration* A^μ can be defined as the *first derivative of four-velocity with respect to proper time*,

$$A^\mu = (A^0, A^1, A^2, A^3) = \frac{dU^\mu}{d\tau}. \qquad (30.11)$$

To express this four-acceleration in terms of three-dimensional velocity and acceleration vectors, we make some rearrangements,

$$A^\mu = \frac{dU^\mu}{dt}\frac{dt}{d\tau} = \gamma_u \frac{dU^\mu}{dt}$$

$$= \gamma_u \left(\frac{d\gamma_u}{dt}(c,\ u_x,\ u_y,\ u_z) + \gamma_u \frac{d}{dt}(c,\ u_x,\ u_y,\ u_z) \right).$$

With (22.3),

$$\frac{d\gamma_u}{dt} = \frac{\gamma_u^3}{c^2}\mathbf{u}\cdot\mathbf{a},$$

we obtain

$$A^\mu = \frac{\gamma_u^4}{c^2}\mathbf{u}\cdot\mathbf{a}(c,\ u_x,\ u_y,\ u_z) + \gamma_u^2 \frac{d}{dt}(c,\ u_x,\ u_y,\ u_z),$$

or in components,

$$A^0 = \frac{\gamma_u^4}{c}\mathbf{u}\cdot\mathbf{a}$$

$$A^1 = \frac{\gamma_u^4}{c^2}\mathbf{u}\cdot\mathbf{a}u_x + \gamma_u^2 a_x$$

$$A^2 = \frac{\gamma_u^4}{c^2}\mathbf{u}\cdot\mathbf{a}u_y + \gamma_u^2 a_y$$

$$A^3 = \frac{\gamma_u^4}{c^2}\mathbf{u}\cdot\mathbf{a}u_z + \gamma_u^2 a_z. \qquad (30.12)$$

These expressions look familiar. When the relativistic law of motion (22.4) is multiplied by γ_u/m_0, one obtains the components A^1, A^2, and A^3 of four-acceleration. The reason for the factor γ_u to appear will be elucidated in Chapter 33. With (30.12), we can now calculate $A^\mu A_\mu$,

$$A^\mu A_\mu = \frac{\gamma_u^8}{c^4}[c^2(\mathbf{u} \cdot \mathbf{a})^2 - (u_x^2 + u_y^2 + u_z^2)(\mathbf{u} \cdot \mathbf{a})^2]$$

$$-2\frac{\gamma_u^6}{c^2}(u_x a_x + u_y a_y + u_z a_z)(\mathbf{u} \cdot \mathbf{a}) - \gamma_u^4(a_x^2 + a_y^2 + a_z^2).$$

With $c^2 - u^2 = c^2(1 - u^2/c^2) = c^2/\gamma_u^2$, we obtain

$$A^\mu A_\mu = \frac{\gamma_u^6}{c^2}(\mathbf{u} \cdot \mathbf{a})^2 - 2\frac{\gamma_u^6}{c^2}(\mathbf{u} \cdot \mathbf{a})^2 - \gamma_u^4 a^2$$

$$A^\mu A_\mu = -\frac{\gamma_u^6}{c^2}(\mathbf{u} \cdot \mathbf{a})^2 - \gamma_u^4 a^2. \tag{30.13}$$

In the instantaneous rest frame, the S'-frame, $\mathbf{u}' = 0$ and $\gamma_{u'} = 1$ holds, thus

$$A'^\mu A'_\mu = -\alpha^2. \tag{30.14}$$

α is the *proper acceleration*. For linear acceleration, i.e. $\mathbf{u} \| \mathbf{a}$, (30.13) and (30.14) give

$$\alpha^2 = \gamma_u^6 \beta_u^2 a^2 + \gamma_u^4 a^2 = \gamma_u^4 a^2 \left(\gamma_u^2 \beta_u^2 + 1\right) = \gamma_u^4 a^2 \left(\frac{\beta_u^2}{1 - \beta_u^2} + 1\right) = \gamma_u^6 a^2$$

$$\alpha = \gamma_u^3 a. \tag{30.15}$$

This result is nothing but (12.3). For circular motion, $\mathbf{u} \perp \mathbf{a}$, one obtains correspondingly

$$\alpha = \gamma_u^2 a. \tag{30.16}$$

In a storage ring, the *centripetal proper acceleration* is $\alpha = \gamma_u^2 \frac{u^2}{r} \approx \gamma_u^2 \frac{c^2}{r}$.

Chapter 31*

The Angular Frequency–Wave Number Vector

A plane wave propagates in the x-direction. It can be described by the equation

$$A(t, x) = A_0 \sin(\omega t - k_x x). \tag{31.1}$$

The *displacement* A is a function of time t and position x, A_0 is the *amplitude*, $\omega = 2\pi/T$ is the *angular frequency* (T is the *period*), and $k_x = 2\pi/\lambda$ is the *wave number* (λ is the *wavelength*). When the plane wave propagates in an arbitrary direction, one writes

$$A(t, x, y, z) = A_0 \sin[\omega t - (k_x x + k_y y + k_z z)],$$

or

$$A(t, \mathbf{r}) = A_0 \sin(\omega t - \mathbf{k} \cdot \mathbf{r}). \tag{31.2}$$

As usual, the *relation between wave number and angular frequency* is

$$|\mathbf{k}| = \frac{2\pi}{\lambda} = \frac{2\pi}{\dfrac{c}{\nu}} = \frac{2\pi\nu}{c} = \frac{\omega}{c}. \tag{31.3}$$

The *argument of the sine function is called the phase*, φ. It indicates whether, at a certain moment of time, t, and position in space, $\mathbf{r}$, the wave is at a maximum, a minimum, or at any intermediate displacement. Quite obviously, it has to be Lorentz invariant. A wave's phase at an event E,

269

whose S-frame position is given by $X^{\mu} = (ct, x, y, z)$, has to have the same value in the S'-frame, given by $X'^{\mu} = (ct', x', y', z')$. Therefore,

$$\varphi = \omega t - (k_x x + k_y y + k_z z) = \omega' t' - (k'_x x' + k'_y y' + k'_z z') = \varphi'.$$

So the angular frequency, as well as the components of the wave number vector, have to have different values $(\omega', k'_x, k'_y, k'_z)$ in the S'-frame. It seems reasonable to write the phase as the scalar product of an *angular frequency–wave number vector*,

$$K^{\mu} = (K^0, K^1, K^2, K^3) = \left(\frac{\omega}{c}, k_x, k_y, k_z \right), \qquad (31.4)$$

and the time-position vector, $X^{\mu} = (ct, x, y, z)$, and to demand that

$$\varphi = K^{\mu} X_{\mu} = K'^{\mu} X'_{\mu} = \varphi' \qquad (31.5)$$

holds.

Of course, the angular frequency–wave number vector introduced above is nothing but the energy-momentum vector: From (25.12) we know that energy is related to frequency by $E = h\nu$. With the abbreviation $\hbar = h/2\pi$, (25.12) can be written as

$$E = \hbar \omega. \qquad (31.6)$$

With (24.4), $E = |\mathbf{p}|c$ is true for photons. With (31.6) and (31.3), we can immediately find the relation $|\mathbf{p}| = \hbar\omega/c = \hbar|\mathbf{k}|$, or in vector notation,

$$\mathbf{p} = \hbar \mathbf{k}. \qquad (31.7)$$

Therefore,

$$P^{\mu} = \hbar K^{\mu}. \qquad (31.8)$$

If (31.5) is true, one should be able to derive K'^{μ} from K^{μ} by means of a Lorentz transformation. To do so, we follow the approach of Chapter 10 (see Fig. 10.1) and assign to **k** a direction in the S-frame of the light receiver given by

$$|\mathbf{k}|(\cos\theta, \sin\theta, 0) = \frac{\omega}{c}(\cos\theta, \sin\theta, 0).$$

Then the angular frequency–wave number vector reads

$$K^\mu = \frac{\omega}{c}(1, \cos\theta, \sin\theta, 0). \tag{31.9}$$

The Lorentz transformation of this vector,

$$
\begin{pmatrix} \omega'/c \\ k'_x \\ k'_y \\ k'_z \end{pmatrix}
=
\begin{pmatrix}
\gamma_v & -\gamma_v\beta_v & 0 & 0 \\
-\gamma_v\beta_v & \gamma_v & 0 & 0 \\
0 & 0 & 1 & 0 \\
0 & 0 & 0 & 1
\end{pmatrix}
\cdot \frac{\omega}{c}
\begin{pmatrix} 1 \\ \cos\theta \\ \sin\theta \\ 0 \end{pmatrix},
$$

gives us the following set of equations:

$$\omega' = \gamma_v\omega(1 - \beta_v\cos\theta) \tag{31.10a}$$

$$k'_x = \gamma_v\frac{\omega}{c}(\cos\theta - \beta_v) \tag{31.10b}$$

$$k'_y = \frac{\omega}{c}\sin\theta \tag{31.10c}$$

$$k'_z = 0. \tag{31.10d}$$

These are the components of the angular frequency–wave number vector in the S'-frame of the light source. If we identify the frequency of the source, ω', with $2\pi\nu_0$ and the frequency of the receiver, ω, with $2\pi\nu$, we obtain from (31.10a) the relation

$$\frac{\nu}{\nu_0} = \frac{1}{\gamma_v(1 - \beta_v\cos\theta)}.$$

This is (14.5), the Doppler equation for a light source passing the light receiver at an arbitrary distance. From (31.10b) follows, with (31.10a),

$$k'_x = \frac{\omega'}{c}\frac{\cos\theta - \beta_v}{1 - \beta_v\cos\theta},$$

and from (31.10c), with (31.10a),

$$k'_y = \frac{\omega'}{c} \frac{\sin \theta}{\gamma_v (1 - \beta_v \cos \theta)}.$$

This gives

$$\frac{k'_y}{k'_x} = \frac{\sin \theta}{\gamma_v (\cos \theta - \beta_v)}. \qquad (31.11)$$

But from Fig. 10.1, one can see that k'_y / k'_x is nothing but

$$\frac{k'_y}{k'_x} = \frac{u'_y}{u'_x} = \tan \theta'. \qquad (31.12)$$

So (31.11) and (31.12) yield the equation of light aberration, (10.3). A Lorentz transformation from K^μ to K'^μ produces reasonable results, indicating that (31.5) is indeed correct.

Chapter 32*

The Charge Density–Current Density Vector

Chapter 5 showed that a transformation of the inertial reference frame may turn a Lorentz force $\mathbf{f}_L$ into a Coulomb force $\mathbf{f}'_C$. For that purpose we introduced the *charge density*

$$\rho = \gamma_u ne = \gamma_u \rho_0. \tag{32.1}$$

ρ_0 is the charge density of a charge distribution at rest. When in motion at the velocity $\mathbf{u}$, this charge density is subjected to length contraction and has to be augmented by a factor of γ_u to give ρ. The unit of charge density is Cm^{-3}.

A charge distribution of charge density ρ moving at the velocity $\mathbf{u}$ gives rise to an *electric current density*

$$\mathbf{j} = \rho \mathbf{u} = \gamma_u \rho_0 \mathbf{u}. \tag{32.2}$$

Its unit is $\text{Cm}^{-2}\text{s}^{-1}$.

Let us tentatively write these four electric quantities as a four-vector, and replace ρ by ρc in order to have the same unit for all components,

$$(\rho c, j_x, j_y, j_z) = \rho(c, u_x, u_y, u_z) = \gamma_u \rho_0 (c, u_x, u_y, u_z).$$

Remembering the energy-momentum vector,

$$\mathbf{P}^\mu = \left(\frac{E}{c}, p_x, p_y, p_z\right) = m(c, u_x, u_y, u_z) = \gamma_u m_0 (c, u_x, u_y, u_z),$$

we find that its four components depend on c and $\mathbf{u}$ in the very same way. This encourages us to define a *charge density–current density vector* by

$$J^\mu = (J^0, J^1, J^2, J^3) = (\rho c, j_x, j_y, j_z) = \rho(c, u_x, u_y, u_z). \qquad (32.3)$$

Equipped with this new four-vector, let us redo the calculation of Chapter 5. In the S-frame,

$$\rho = \rho_+ + \rho_- = 0 \qquad (\text{i.e. } \rho_+ = -\rho_-)$$

$$j_x = \rho_- v$$

$$j_y = 0$$

$$j_z = 0$$

holds. v is the velocity of the electrons and, at the same time, the relative velocity between the S-frame and the S'-frame. Then, a Lorentz transformation gives us

$$\begin{pmatrix} \rho'c \\ j'_x \\ j'_y \\ j'_z \end{pmatrix} = \begin{pmatrix} \gamma_v & -\gamma_v\beta_v & 0 & 0 \\ -\gamma_v\beta_v & \gamma_v & 0 & 0 \\ 0 & 0 & 1 & 0 \\ 0 & 0 & 0 & 1 \end{pmatrix} \begin{pmatrix} 0 \\ \rho_- v \\ 0 \\ 0 \end{pmatrix} = \begin{pmatrix} -\gamma_v\beta_v\rho_- v \\ \gamma_v\rho_- v \\ 0 \\ 0 \end{pmatrix},$$

$$(32.4)$$

or

$$\rho' = -\gamma_v\beta_v^2\rho_- = \gamma_v\beta_v^2\rho_+ \qquad (32.5a)$$

$$j'_x = \gamma_v\rho_- v = -\gamma_v\rho_+ v. \qquad (32.5b)$$

ρ' is identical with the charge density derived in (5.2). j'_x was not used in Chapter 5, but is self-explanatory: Due to the length contraction the charge density of a positive crystal lattice increases by a factor of γ_v from ρ_+ to $\gamma_v\rho_+$. Then the motion of the crystal lattice at the velocity $-v$ gives rise to a current density of $-\gamma_v\rho_+ v$.

Chapter 33*

The Power-Force Vector (Minkowski Force)

Even the three-dimensional force vector **f** can be expanded into a four-dimensional vector. Unfortunately, the upper-case letter F, in bold-face though, is reserved for the electromagnetic field-strength tensor (Chapter 34); therefore we switch to K.

The *four-dimensional force vector K^μ*, also called *Minkowski force*, is the *derivative of the energy-momentum vector P^μ with respect to the Lorentz invariant proper time τ,*

$$K^\mu = \frac{dP^\mu}{d\tau}. \tag{33.1}$$

$d\tau = dt/\gamma_u$, with γ_u given by (22.4), yields

$$K^\mu = \gamma_u \left(\frac{d}{dt}\frac{E}{c}, \frac{dp_x}{dt}, \frac{dp_y}{dt}, \frac{dp_z}{dt} \right). \tag{33.2}$$

With (23.2), (23.1), and (23.6), we obtain

$$\frac{d}{dt}E = \frac{d}{dt}(m_0 c^2 + E_{kin}) = \frac{d}{dt}E_{kin} = \mathbf{f} \cdot \mathbf{u}.$$

This equation, together with (22.2), turn (33.2) into

$$K^\mu = (K^0, K^1, K^2, K^3) = \gamma_u \left(\frac{1}{c}\mathbf{f} \cdot \mathbf{u}, f_x, f_y, f_z \right). \tag{33.3}$$

$\mathbf{u} = (u_x, u_y, u_z)$ is the velocity of an object acted upon by a force $\mathbf{f}$. K^μ is known as the *power-force vector*; its zeroth component is power divided by c, leaving the first, second, and third components as force components. Now we can also tentatively *generalize Newton's second law*, $\mathbf{f} = m_0\mathbf{a}$, and write

$$K^\mu = m_0 A^\mu. \tag{33.4}$$

We prove (33.4) component by component with (33.3), (30.12), and (22.6) for $\mu = 0$:

$$K^0 \overset{!}{=} m_0 A^0$$

$$\gamma_u \frac{1}{c}\mathbf{f}\cdot\mathbf{u} = m_0\frac{\gamma_u^4}{c}\mathbf{u}\cdot\mathbf{a} = m_0\frac{\gamma_u^4}{c}\frac{1}{\gamma_u m_0}\mathbf{f}\cdot\mathbf{u}\left(1-\frac{u^2}{c^2}\right) = \gamma_u\frac{1}{c}\mathbf{f}\cdot\mathbf{u},$$

Q.E.D., and with (33.3), (30.12), and (22.4) for $\mu = 1, 2, 3$:

$$K^{1,2,3} \overset{!}{=} m_0 A^{1,2,3}$$

$$\gamma_u\mathbf{f} = m_0\left[\frac{\gamma_u^4}{c^2}(\mathbf{u}\cdot\mathbf{a})\mathbf{u} + \gamma_u^2\mathbf{a}\right] = \gamma_u\mathbf{f},$$

Q.E.D.

With (33.3), one can *transform a force $\mathbf{f}$ of the S-frame into a corresponding force $\mathbf{f}'$ of the S'-frame*. But to do so, one has to transform K^μ. We show this by using the Lorentz transformation for the x-direction,

$$K'^0 = \gamma_v(K^0 - \beta_v K^1) \tag{33.5a}$$

$$K'^1 = \gamma_v(K^1 - \beta_v K^0) \tag{33.5b}$$

$$K'^2 = K^2 \tag{33.5c}$$

$$K'^3 = K^3. \tag{33.5d}$$

We use for the left-hand and right-hand sides of (33.5 b, c, d) the expression (33.3) for primed and unprimed variables, respectively, and solve for $f'_x, f'_y,$ and $f'_z,$

$$\gamma_{u'} f'_x = \gamma_v \left(\gamma_u f_x - \beta_v \frac{\gamma_u}{c} \mathbf{f} \cdot \mathbf{u} \right)$$

$$\gamma_{u'} f'_y = \gamma_u f_y$$

$$\gamma_{u'} f'_z = \gamma_u f_z.$$

For $\gamma_{u'}$, we take (30.5) and obtain

$$f'_x = \frac{f_x - \dfrac{v}{c^2} \mathbf{f} \cdot \mathbf{u}}{1 - \dfrac{u_x v}{c^2}} \qquad (33.6a)$$

$$f'_y = \frac{f_y}{\gamma_v \left(1 - \dfrac{u_x v}{c^2} \right)} \qquad (33.6b)$$

$$f'_z = \frac{f_z}{\gamma_v \left(1 - \dfrac{u_x v}{c^2} \right)}. \qquad (33.6c)$$

For the inverse transformation one exchanges the variables $\mathbf{u}$ and $\mathbf{f}$ (more precisely, their components) by $\mathbf{u}'$ and $\mathbf{f}'$, and v by $-v$, as usual. If the object, acted upon by the force under consideration, is at rest in the S-frame, $\mathbf{u} = 0$ holds, and (33.6) is simplified to

$$f'_x = f_x \qquad (33.7a)$$

$$f'_y = \frac{1}{\gamma_v} f_y \qquad (33.7b)$$

$$f'_z = \frac{1}{\gamma_v} f_z. \qquad (33.7c)$$

We now want to illustrate the transformation of forces by redoing the calculation of Chapter 5 where we showed that the Lorentz force and the Coulomb force are nothing but two instances of one and the same force which, depending on the inertial reference frame chosen, appears in one or the other form or in a combination of both. (5.9) says that the transverse Coulomb force in the S'-frame, f_C', is γ_v times as large as the Lorentz force, f_L, in the S-frame: $f_C' = \gamma_v f_L$. This result is now immediately obvious from (33.6c). The charge which is acted upon by the force moves at $u_x = v$; thus $f_z' = \gamma_v f_z$.

Chapter 34*

Transformation of Electric and Magnetic Fields

A great advantage of the power-force vector is that it enables us to derive a solution for the *Lorentz transformation of the electric field*, E, *and the magnetic flux density, or magnetic induction*, B. We start with the equation

$$\mathbf{f} = q(\mathbf{E} + \mathbf{u} \times \mathbf{B}). \tag{34.1}$$

The first summand, $q\mathbf{E}$, describes the *electric force*, the second one, $q(\mathbf{u} \times \mathbf{B})$, gives the *magnetic force* on a charge q moving at the velocity $\mathbf{u}$. That (34.1) applies in a relativistic context is a *working hypothesis of the special theory of relativity*, as pointed out in Chapter 1. This hypothesis is the foundation of the transformation equations for E and B. With (33.3), the four-vector pertaining to (34.1) reads

$$K^{\mu} = \gamma_u q \left(\frac{1}{c} \mathbf{E} \cdot \mathbf{u}, \mathbf{E} + \mathbf{u} \times \mathbf{B} \right), \tag{34.2}$$

as the scalar triple product $(\mathbf{u} \times \mathbf{B}) \cdot \mathbf{u}$ is zero. Written component by component, we get

$$\begin{pmatrix} K^0 \\ K^1 \\ K^2 \\ K^3 \end{pmatrix} = \gamma_u q \begin{pmatrix} E_x u_x/c + E_y u_y/c + E_z u_z/c \\ E_x + u_y B_z - u_z B_y \\ E_y + u_z B_x - u_x B_z \\ E_z + u_x B_y - u_y B_x \end{pmatrix},$$

or as a matrix equation,

$$\begin{pmatrix} K^0 \\ K^1 \\ K^2 \\ K^3 \end{pmatrix} = \frac{q}{c} \begin{pmatrix} 0 & E_x & E_y & E_z \\ E_x & 0 & cB_z & -cB_y \\ E_y & -cB_z & 0 & cB_x \\ E_z & cB_y & -cB_x & 0 \end{pmatrix} \begin{pmatrix} \gamma_u c \\ \gamma_u u_x \\ \gamma_u u_y \\ \gamma_u u_z \end{pmatrix}, \qquad (34.3)$$

which can be written more concisely as

$$K^\mu = \frac{q}{c} F U^\mu. \qquad (34.4)$$

F is called the *electromagnetic field-strength tensor*, or short *field-strength tensor*. At this point, it is customary to introduce the *tensor formalism*, replacing the boldface notation (**F**) with two superscripts ($F^{\mu\nu}$), or two subscripts ($F_{\mu\nu}$), or one superscript and one subscript (F^ν_μ), always in Greek. Since we have now almost completed our course in the special theory of relativity, we will dispense with this formalism, noting, however, that it will be indispensable in the general theory of relativity.

We now write down in matrix form the Lorentz transformation, **L**, which transforms the S-frame to the S'-frame with $\mathbf{v} = (v, 0, 0)$, and the inverse Lorentz transformation, $\mathbf{L}^{-1}$, which transforms the S'-frame to the S-frame with $-\mathbf{v} = (-v, 0, 0)$,

$$\mathbf{L} = \begin{pmatrix} \gamma_v & -\gamma_v \beta_v & 0 & 0 \\ -\gamma_v \beta_v & \gamma_v & 0 & 0 \\ 0 & 0 & 1 & 0 \\ 0 & 0 & 0 & 1 \end{pmatrix} \quad \mathbf{L}^{-1} = \begin{pmatrix} \gamma_v & \gamma_v \beta_v & 0 & 0 \\ \gamma_v \beta_v & \gamma_v & 0 & 0 \\ 0 & 0 & 1 & 0 \\ 0 & 0 & 0 & 1 \end{pmatrix}. \qquad (34.5)$$

One can quickly prove that the *unit matrix* is formed by multiplying **L** and $\mathbf{L}^{-1}$,

$$\mathbf{L}\mathbf{L}^{-1} = \mathbf{L}^{-1}\mathbf{L} = \mathbf{1}. \qquad (34.6)$$

This enables us to write the transformation of the Minkowski force as

$$K'^\mu = L K^\mu, \quad K^\mu = L^{-1} K'^\mu. \tag{34.7}$$

Any other four-vector can be transformed in an analogous way.

Let us now return to (34.4). Since the charge, q, and the speed of light, c, are invariant, we can write

$$K'^\mu = \frac{q}{c} F' U'^\mu. \tag{34.8}$$

F' is the electromagnetic field-strength tensor in the S'-frame. Then it follows from (34.8), (34.7), and (34.4) that

$$\frac{q}{c} F' U'^\mu = L \frac{q}{c} F U^\mu,$$

and, with (34.6),

$$F' U'^\mu = L F L^{-1} L U^\mu$$
$$F' U'^\mu = L F L^{-1} U'^\mu$$

$$F' = L F L^{-1}. \tag{34.9}$$

Written in components,

$$
\begin{pmatrix}
0 & E'_x & E'_y & E'_z \\
E'_x & 0 & cB'_z & -cB'_y \\
E'_y & -cB'_z & 0 & cB'_x \\
E'_z & cB'_y & -cB'_x & 0
\end{pmatrix}
=
\begin{pmatrix}
\gamma_v & -\gamma_v \beta_v & 0 & 0 \\
-\gamma_v \beta_v & \gamma_v & 0 & 0 \\
0 & 0 & 1 & 0 \\
0 & 0 & 0 & 1
\end{pmatrix}
$$

$$
\times
\begin{pmatrix}
0 & E_x & E_y & E_z \\
E_x & 0 & cB_z & -cB_y \\
E_y & -cB_z & 0 & cB_x \\
E_z & cB_y & -cB_x & 0
\end{pmatrix}
\begin{pmatrix}
\gamma_v & \gamma_v \beta_v & 0 & 0 \\
\gamma_v \beta_v & \gamma_v & 0 & 0 \\
0 & 0 & 1 & 0 \\
0 & 0 & 0 & 1
\end{pmatrix}.
$$

The $\times$ indicates that the matrix multiplication has to be continued on the next line, the order of multiplication is irrelevant. Doing so, the right-hand side becomes

$$\begin{pmatrix} 0 & E_x & \gamma_v(E_y - vB_z) & \gamma_v(E_z + vB_y) \\ E_x & 0 & c\gamma_v\left(B_z - \frac{v}{c^2}E_y\right) & -c\gamma_v\left(B_y + \frac{v}{c^2}E_z\right) \\ \gamma_v(E_y - vB_z) & -c\gamma_v\left(B_z - \frac{v}{c^2}E_y\right) & 0 & cB_x \\ \gamma_v(E_z + vB_y) & c\gamma_v\left(B_y + \frac{v}{c^2}E_z\right) & -cB_x & 0 \end{pmatrix}.$$

A comparison component by component yields *the Lorentz transformations of* **E** *and* **B**,

$$E'_x = E_x \qquad\qquad B'_x = B_x$$

$$E'_y = \gamma_v(E_y - vB_z) \qquad B'_y = \gamma_v\left(B_y + \frac{v}{c^2}E_z\right)$$

$$E'_z = \gamma_v(E_z + vB_y) \qquad B'_z = \gamma_v\left(B_z - \frac{v}{c^2}E_y\right). \qquad (34.10)$$

We would now like to present *two examples* of how to use the Lorentz transformation of the E-field and B-field.

Let us first return to Chapter 5, namely *the transformation of a magnetic (Lorentz) force into an electric (Coulomb) force* (Fig. 34.1). In that calculation, $\mathbf{v} = (v, 0, 0)$ was both the relative velocity between the S-frame

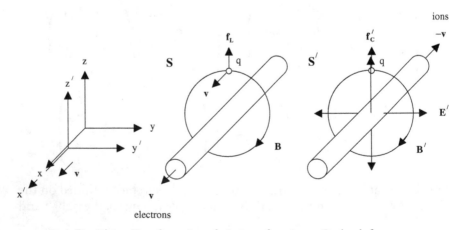

Fig. 34.1. Transformation of a Lorentz force into a Coulomb force.

and the S'-frame, and the velocity of the conduction electrons inside the copper filament, and also the velocity of the charge q directly above the filament.

At the position of charge q, i.e. above the filament, we can write in the S-frame

$$\mathbf{E} = (0, 0, 0)$$

$$\mathbf{B} = (0, B_y, 0).$$

This gives for the Coulomb force and the Lorentz force

$$\mathbf{f}_C = (0, 0, 0)$$

$$\mathbf{f}_L = (0, 0, qvB_y).$$

With (34.10) we write in the S'-frame

$$\mathbf{E}' = (0, 0, \gamma_v vB_y)$$

$$\mathbf{B}' = (0, \gamma_v B_y, 0).$$

So the Coulomb and Lorentz forces are

$$\mathbf{f}'_C = q\mathbf{E}' = (0, 0, \gamma_v qvB_y) \tag{34.11}$$

$$\mathbf{f}'_L = (0, 0, 0), \tag{34.12}$$

as in the S'-frame the velocity of the charge is zero. The Lorentz force $\mathbf{f}_L$ is thus turned into a Coulomb force $\mathbf{f}'_C$, augmented by a factor of γ_v.

We can explain the γ-factor from the definition of force as the time derivative of momentum: $d\mathbf{p} = \mathbf{f}\, dt$ in the S-frame, and $d\mathbf{p}' = \mathbf{f}'\, dt'$ in the S'-frame. In our problem, the momentum vector increments, $d\mathbf{p}$ and $d\mathbf{p}'$, are perpendicular to the relative velocity $\mathbf{v}$ between the S- and the S'-frames. In Chapter 20, we postulated that transverse momentum components are invariant, $d\mathbf{p} = d\mathbf{p}'$. This leaves us with the simple equation $f\, dt = f'\, dt'$. Knowing that time passes more slowly in the S'-frame than in the S-frame by a factor of γ_v, we can immediately write $dt' = dt/\gamma_v$. That is why the force in the S'-frame has to be larger by a factor of γ_v which yields (34.11), $f' = \gamma_v f$.

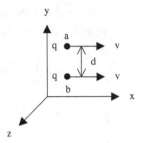

Fig. 34.2. Two particles of charge q having a mutual distance of d and moving at the velocity v parallel to the x-axis.

In a *second example*, we have two particles in the S-frame, a and b, each one carrying a charge q, moving at the mutual distance d parallel to the x-axis at a velocity of $\mathbf{v} = (v, 0, 0)$ (Fig. 34.2). What is the magnitude of the force between the two charges?

For this problem we can offer two independent solution strategies. The first and more difficult track starts out in the S'-frame moving, as usual, at a velocity of $\mathbf{v} = (v, 0, 0)$ relative to the S-frame. In the S'-frame, a and b are at rest which, for each charge, makes the magnetic field of the other charge disappear. Thus, at the position of a,

$$\mathbf{E}' = \left(0, \frac{1}{4\pi\varepsilon_0}\frac{q}{d^2}, 0\right)$$

$$\mathbf{B}' = (0, 0, 0).$$

With (34.10), the inverse Lorentz transformation is

$$\mathbf{E} = \left(0, \gamma_v \frac{1}{4\pi\varepsilon_0}\frac{q}{d^2}, 0\right)$$

$$\mathbf{B} = \left(0, 0, \gamma_v \frac{v}{c^2}\frac{1}{4\pi\varepsilon_0}\frac{q}{d^2}\right).$$

In the S-frame, the force on particle a becomes

$$\mathbf{f} = q(\mathbf{E} + \mathbf{v} \times \mathbf{B})$$

$$= \left(0, \gamma_v \frac{1}{4\pi\varepsilon_0}\frac{q^2}{d^2} - \gamma_v \beta_v^2 \frac{1}{4\pi\varepsilon_0}\frac{q^2}{d^2}, 0\right) = \left(0, (1 - \beta_v^2)\gamma_v \frac{1}{4\pi\varepsilon_0}\frac{q^2}{d^2}, 0\right),$$

or

$$\mathbf{f} = \left(0, \frac{1}{\gamma_v}\frac{1}{4\pi\varepsilon_0}\frac{q^2}{d^2}, 0\right). \tag{34.13}$$

In the S-frame, the force is smaller by a factor of γ_v when compared with the S'-frame.

The second approach invokes the power-force vector (33.3) and starts in the S'-frame. The velocity $\mathbf{u}'$ of both particles is zero there, thus $\gamma_{u'} = 1$. The force at a is then given by

$$K'^{\mu} = (K'^0, K'^1, K'^2, K'^3) = \left(0, 0, \frac{1}{4\pi\varepsilon_0}\frac{q^2}{d^2}, 0\right).$$

The inverse Lorentz transformation gives, with the help of (33.5),

$$K^{\mu} = (K^0, K^1, K^2, K^3) = \left(0, 0, \frac{1}{4\pi\varepsilon_0}\frac{q^2}{d^2}, 0\right).$$

With $K^2 = \gamma_v f_y$, and noting that $\gamma_v \neq 1$ as both particles are in motion in the S-frame, we obtain

$$f_y = \frac{1}{\gamma_v}\frac{1}{4\pi\varepsilon_0}\frac{q^2}{d^2},$$

and thus the result (34.13).

Finally, we would like to derive Lorentz transformations of the E-field and B-field for the more general case of a relative velocity of $\mathbf{v} = (v_x, v_y, v_z)$ between the S-frame and the S'-frame. We begin by rewriting (34.10), i.e. the field components for the transformation with $\mathbf{v} = (v, 0, 0)$,

$$E'_x = E_x \qquad\qquad B'_x = B_x$$

$$E'_y = \gamma_v[E_y + (\mathbf{v} \times \mathbf{B})_y] \qquad B'_y = \gamma_v\left[B_y - \frac{1}{c^2}(\mathbf{v} \times \mathbf{E})_y\right]$$

$$E'_z = \gamma_v[E_z + (\mathbf{v} \times \mathbf{B})_z] \qquad B'_z = \gamma_v\left[B_z - \frac{1}{c^2}(\mathbf{v} \times \mathbf{E})_z\right]. \tag{34.14}$$

One identifies the first line as *field components parallel to* $\mathbf{v}$, and the second and third lines as *field components perpendicular to* $\mathbf{v}$. We introduce the subscripts $\parallel$ and $\perp$, and write

$$E'_\parallel = E_\parallel \qquad\qquad\qquad B'_\parallel = B_\parallel$$

$$E'_\perp = \gamma_v[E_\perp + (v \times B)_\perp] \qquad B'_\perp = \gamma_v\left[B_\perp - \frac{1}{c^2}(v \times E)_\perp\right]$$

$$= \gamma_v(E_\perp + v \times B) \qquad = \gamma_v\left(B_\perp - \frac{1}{c^2}v \times E\right). \qquad (34.15)$$

The last conversion is possible as a cross product is always a vector pointing at right angles to the vectors from which it is formed, which makes one of the indices $\perp$ superfluous. (34.15) now holds for any arbitrary v; the limitation to $v = (v, 0, 0)$ is overcome. Following the prescriptions of (18.4) and (18.5), we write $E_\parallel$ and $E_\perp$ as projections of E on v,

$$E_\parallel = \frac{(v \cdot E)v}{v^2} \qquad\qquad (34.16)$$

$$E_\perp = E - E_\parallel = E - \frac{(v \cdot E)v}{v^2}. \qquad\qquad (34.17)$$

Corresponding results hold for $B_\parallel$ and $B_\perp$. This enables us to write

$$E' = E'_\parallel + E'_\perp = \frac{(v \cdot E)v}{v^2} + \gamma_v\left[E - \frac{(v \cdot E)v}{v^2} + v \times B\right]$$

$$B' = B'_\parallel + B'_\perp = \frac{(v \cdot B)v}{v^2} + \gamma_v\left[B - \frac{(v \cdot B)v}{v^2} - \frac{1}{c^2}v \times E\right]$$

$$E' = \gamma_v(E + v \times B) - (\gamma_v - 1)\frac{(v \cdot E)v}{v^2}$$

$$B' = \gamma_v\left(B - \frac{1}{c^2}v \times E\right) - (\gamma_v - 1)\frac{(v \cdot B)v}{v^2}. \qquad (34.18)$$

For $v/c \to 0$, we obtain the non-relativistic approximation,

$$E' = E + v \times B$$

$$B' = B - \frac{1}{c^2}v \times E. \qquad (34.19)$$

It is interesting to note that the *electric and magnetic fields intermingle at any velocity, no matter how small*. This astonishes us so much more as the appearance of the $\mathbf{E}'$-field in the S'-frame is due to length contraction, as was explored in Chapter 5.

From the $\mathbf{E}$- and $\mathbf{B}$-fields, two *Lorentz invariant quantities* can be formed which are called the *invariants of the electromagnetic field*,

$$\mathbf{E}' \cdot \mathbf{B}' = \mathbf{E} \cdot \mathbf{B} \qquad (34.20)$$

$$E'^2 - c^2 B'^2 = E^2 - c^2 B^2. \qquad (34.21)$$

Without a loss of generality, we can prove (34.20) and (34.21) with (34.10):

$$\mathbf{E}' \cdot \mathbf{B}' = E'_x B'_x + E'_y B'_y + E'_z B'_z$$

$$= E_x B_x + \gamma_v^2 \left(E_y B_y + \frac{v}{c^2} E_y E_z - v B_y B_z - \frac{v^2}{c^2} E_z B_z \right)$$

$$+ \gamma_v^2 \left(E_z B_z - \frac{v}{c^2} E_y E_z + v B_y B_z - \frac{v^2}{c^2} E_y B_y \right)$$

$$= E_x B_x + \gamma_v^2 \left[E_y B_y \left(1 - \frac{v^2}{c^2} \right) + E_z B_z \left(1 - \frac{v^2}{c^2} \right) \right]$$

$$= E_x B_x + E_y B_y + E_z B_z = \mathbf{E} \cdot \mathbf{B},$$

Q.E.D., and

$$E'^2 - c^2 B'^2 = E'^2_x + E'^2_y + E'^2_z - c^2 B'^2_x - c^2 B'^2_y - c^2 B'^2_z$$

$$= E^2_x + \gamma_v^2 (E^2_y - 2v E_y B_z + v^2 B^2_z + E^2_z + 2v E_z B_y + v^2 B^2_y)$$

$$- c^2 B^2_x - c^2 \gamma_v^2 \left(B^2_y + 2\frac{v}{c^2} E_z B_y + \frac{v^2}{c^4} E^2_z + B^2_z - 2\frac{v}{c^2} E_y B_z + \frac{v^2}{c^4} E^2_y \right)$$

$$= E^2_x - c^2 B^2_x + \gamma_v^2 \left[E^2_y \left(1 - \frac{v^2}{c^2} \right) + E^2_z \left(1 - \frac{v^2}{c^2} \right) \right.$$

$$\left. + B^2_y (v^2 - c^2) + B^2_z (v^2 - c^2) \right]$$

$$= E^2_x - c^2 B^2_x + E^2_y + E^2_z - c^2 B^2_y - c^2 B^2_z = E^2 - c^2 B^2,$$

Q.E.D.

These somewhat tedious calculations enable us to infer some interesting consequences, namely:

- If **E** and **B** are at right angles to each other ($\mathbf{E} \cdot \mathbf{B} = 0$), they are so in all inertial reference frames. Electromagnetic radiation serves as an example.

- If **E** and **B** form an acute angle or an obtuse angle, i.e. if $\mathbf{E} \cdot \mathbf{B} = |\mathbf{E}||\mathbf{B}| \cos\varphi > 0$ or $\mathbf{E} \cdot \mathbf{B} = |\mathbf{E}||\mathbf{B}| \cos\varphi < 0$, respectively, they do so in all other inertial reference frames. A transformation into reference frames with $\mathbf{E} = 0$ and/or $\mathbf{B} = 0$ is impossible. In other words: If a Lorentz transformation is to turn **E** or **B** into zero, **E** and **B** have to stand at right angles to each other.

- If $|\mathbf{E}| = c|\mathbf{B}|$, or $E^2 - c^2 B^2 = 0$, then this is also true for all inertial reference frames. Electromagnetic radiation may again be cited as an example.

- If there is an E-field but no B-field in the S-frame, a transformation to a pure B'-field in the S'-frame is impossible and vice versa. If it were possible, we would have

$$-c^2 B'^2 = E^2,$$

or

$$E'^2 = -c^2 B^2,$$

which, because of the sign, can only be true for a vanishing E-field or B-field.

- If in any inertial frame $|\mathbf{E}| > c|\mathbf{B}|$ holds, i.e. $E^2 - c^2 B^2 > 0$, then this must also be true for any other inertial reference frame. Then, as shown before, the fields can be transformed to a pure E-field if $\mathbf{E} \cdot \mathbf{B} = 0$ holds. Correspondingly, for $E^2 - c^2 B^2 < 0$, the fields can be transformed to a pure B-field for $\mathbf{E} \cdot \mathbf{B} = 0$.

Chapter 35*

Fields of a Passing Point Charge

This chapter follows closely Jackson's treatment of the fields of a passing point charge. A charge q is at rest in the origin of the S'-frame which moves at the velocity $\mathbf{v} = (v, 0, 0)$ relative to the S-frame. At the time $t = t' = 0$, the origins of both frames coincide (Fig. 35.1). An observer is at rest in the S-frame at the point P ($x = 0 \mid y = a \mid z = 0$); his coordinates in the S'-frame are P' ($x' = -vt' \mid y' = a \mid z' = 0$).

What are the electric and magnetic fields caused by the charge as a function of time at point P? In the S'-frame, the electric field has the magnitude

$$|E'| = \frac{1}{4\pi\varepsilon_0} \frac{q}{r'^2} = \frac{1}{4\pi\varepsilon_0} \frac{q}{x'^2 + y'^2 + z'^2}.$$

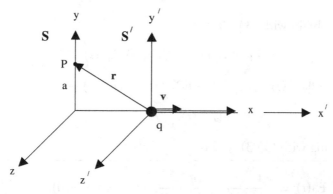

Fig. 35.1. An observer of the S-frame at point P is watching a charge q pass him.

In vector notation,

$$\mathbf{E}' = (E'_x, E'_y, E'_z) = \frac{1}{4\pi\varepsilon_0} \frac{q}{r'^3} (x', y', z'),$$

or

$$\mathbf{E}' = \frac{1}{4\pi\varepsilon_0} \frac{q}{(x'^2 + y'^2 + z'^2)^{3/2}} (x', y', z'). \tag{35.1}$$

As a function of time, the electric field at P is

$$\mathbf{E}'_P(t') = \frac{1}{4\pi\varepsilon_0} \frac{q}{(v^2 t'^2 + a^2)^{3/2}} (-vt', a, 0). \tag{35.2}$$

We now have to subject this formula to *two transformations*: with (6.5), a transformation of the time variable, $t' \to t$, and with (34.10), a transformation of the electric field, $\mathbf{E}'_P \to \mathbf{E}_P$. Since $x = 0$, we write for P

$$t' = \gamma_v \left(t - \frac{v}{c^2} x \right) = \gamma_v t,$$

and obtain

$$\mathbf{E}'_P(t) = \frac{1}{4\pi\varepsilon_0} \frac{q}{(\gamma_v^2 v^2 t^2 + a^2)^{3/2}} (-\gamma_v vt, a, 0). \tag{35.3}$$

Before we are able to transform the electric field, we first have to write down the magnetic induction in the S'-frame, noting that the charge q is at rest there,

$$\mathbf{B}' = \mathbf{B}'_P(t') = (0, 0, 0) = \mathbf{B}'_P(t). \tag{35.4}$$

Thus, we obtain with (34.10)

$$\mathbf{E}_P(t) = (E_{Px}(t), E_{Py}(t), E_{Pz}(t))$$

$$= [E'_{Px}(t), \gamma_v(E'_{Py}(t) + vB'_{Pz}(t)), \gamma_v(E'_{Pz}(t) - vB'_{Py}(t))]$$

$$= (E'_{Px}(t), \gamma_v E'_{Py}(t), 0),$$

which, along with (35.3), yields

$$\mathbf{E}_P(t) = \frac{1}{4\pi\varepsilon_0} \frac{q}{(\gamma_v^2 v^2 t^2 + a^2)^{3/2}} (-\gamma_v vt, \gamma_v a, 0). \tag{35.5}$$

For the magnetic induction we write down

$$\mathbf{B}_P(t) = (B_{Px}(t), B_{Py}(t), B_{Pz}(t))$$

$$= \left[B'_{Px}(t),\ \gamma_v\left(B'_{Py}(t) - \frac{v}{c^2}E'_{Pz}(t)\right),\ \gamma_v\left(B'_{Pz}(t) + \frac{v}{c^2}E'_{Py}(t)\right) \right]$$

$$= \left(0, 0, \gamma_v\frac{v}{c^2}E'_{Py}(t)\right),$$

and with (35.3),

$$\mathbf{B}_P(t) = \frac{1}{4\pi\varepsilon_0} \frac{q}{\left(\gamma_v^2 v^2 t^2 + a^2\right)^{3/2}} \left(0, 0, \gamma_v\frac{v}{c^2}a\right), \tag{35.6}$$

or, by simply comparing (35.5) and (35.6),

$$\mathbf{B}_P(t) = \left(0, 0, \frac{v}{c^2}E_{Py}(t)\right). \tag{35.7}$$

Let us take a look at (35.5). One immediately finds that $E_{Px}/E_{Py} = -vt/a$.

The components of the E-field at P and of the unit vector from q to P form the same ratio. Therefore, the E-field is radial, as is the case for a charge at rest. This result is by no means trivial. After all, the electric field propagates at a finite speed, the speed of light. Consequently, one might presume that the direction of the E-field at P is not determined by the instantaneous position of the charge, but by previous positions; one might expect curved electric field lines.

Two special cases of (35.5) are of interest: When P is *directly above q*, i.e. at $t = 0$, then

$$\mathbf{E}_P(t = 0) = \left(0, \gamma_v\frac{1}{4\pi\varepsilon_0}\frac{q}{a^2}, 0\right). \tag{35.8}$$

The electric field has only a y-component and is larger than the corresponding static field by a factor of γ_v. This result has already been found in the second example of Chapter 34. When P is *directly in front of or behind q*, i.e. when $a = 0$, then

$$\mathbf{E}_P(t) = \left(-\frac{1}{4\pi\varepsilon_0}\frac{vt}{\gamma_v^2(v^2t^2)^{3/2}}, 0, 0\right). \tag{35.9}$$

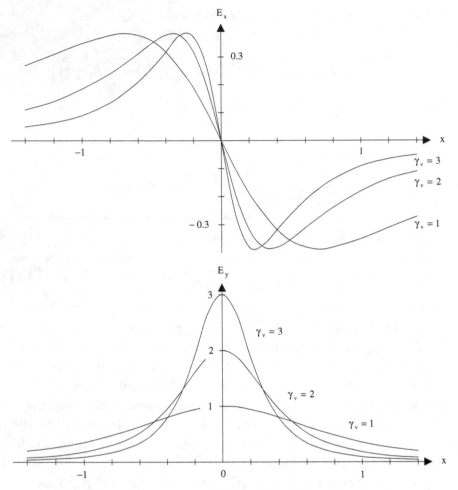

Fig. 35.2.   *x*- and *y*-components of the electric field of a passing point charge.

Note that $vt/(v^2t^2)^{3/2} \neq 1/v^2t^2$ for $t < 0$. Now, the electric field has only an *x*-component and is smaller by a factor of γ_v^2 than the static field.

Figure 35.2 shows (35.5). For simplicity, we set $\frac{1}{4\pi\varepsilon_0} = 1$, $q = 1$, $vt = x$, and $a = 1$. The only parameter left is γ_v. One can see that the longitudinal and the transverse field components decrease with increasing velocity when the charge is still distant. When it approaches the origin, however, the field components increase with velocity.

Chapter 36*

The Shape Invariance of Maxwell's Equations

With the transformation formulas for **E** and **B**, given by (34.10), we can recapitulate one of the most central proofs of theoretical physics, namely the *demonstration of shape invariance of Maxwell's equations*. It is remarkable that shape invariance is already included in Maxwell's equations as they were formulated decades before the special theory of relativity.

To show this proof in detail, we have to delve into the mathematics. Let us start with a function f that depends on the variables ct', x', y', and z',

$$f = f(ct', x', y', z'). \qquad (36.1)$$

If these four variables depend on a new set of variables, ct, x, y, and z,

$$\begin{aligned}
ct' &= ct'(ct, x, y, z) \\
x' &= x'(ct, x, y, z) \\
y' &= y'(ct, x, y, z) \\
z' &= z'(ct, x, y, z),
\end{aligned}$$

and are continuously differentiable with respect to ct, x, y, and z, we can write for the partial derivatives of the function f with respect to ct, x, y, and z,

$$\frac{\partial f}{c\partial t} = \frac{\partial f}{c\partial t'}\frac{c\partial t'}{c\partial t} + \frac{\partial f}{\partial x'}\frac{\partial x'}{c\partial t} + \frac{\partial f}{\partial y'}\frac{\partial y'}{c\partial t} + \frac{\partial f}{\partial z'}\frac{\partial z'}{c\partial t}, \qquad (36.2)$$

and corresponding expressions for $\partial f/\partial x$, $\partial f/\partial y$, and $\partial f/\partial z$. One summarizes these four lines with the *Jacobi matrix*, expressing the differentials

293

of the new coordinate frame (unprimed) as a function of the differentials of the old coordinate frame (primed),

$$
\begin{pmatrix} \dfrac{\partial}{c\partial t} \\[2ex] \dfrac{\partial}{\partial x} \\[2ex] \dfrac{\partial}{\partial y} \\[2ex] \dfrac{\partial}{\partial z} \end{pmatrix}
=
\begin{pmatrix}
\dfrac{c\partial t'}{c\partial t} & \dfrac{\partial x'}{c\partial t} & \dfrac{\partial y'}{c\partial t} & \dfrac{\partial z'}{c\partial t} \\[2ex]
\dfrac{c\partial t'}{\partial x} & \dfrac{\partial x'}{\partial x} & \dfrac{\partial y'}{\partial x} & \dfrac{\partial z'}{\partial x} \\[2ex]
\dfrac{c\partial t'}{\partial y} & \dfrac{\partial x'}{\partial y} & \dfrac{\partial y'}{\partial y} & \dfrac{\partial z'}{\partial y} \\[2ex]
\dfrac{c\partial t'}{\partial z} & \dfrac{\partial x'}{\partial z} & \dfrac{\partial y'}{\partial z} & \dfrac{\partial z'}{\partial z}
\end{pmatrix}
\begin{pmatrix} \dfrac{\partial}{c\partial t'} \\[2ex] \dfrac{\partial}{\partial x'} \\[2ex] \dfrac{\partial}{\partial y'} \\[2ex] \dfrac{\partial}{\partial z'} \end{pmatrix} .
\tag{36.3}
$$

In what follows we restrict ourselves again to the special case that the S'-frame and the S-frame are connected by a transformation with $\mathbf{v} = (v, 0, 0)$, or

$$
\begin{pmatrix} ct' \\ x' \\ y' \\ z' \end{pmatrix}
=
\begin{pmatrix}
\gamma & -\gamma\beta & 0 & 0 \\
-\gamma\beta & \gamma & 0 & 0 \\
0 & 0 & 1 & 0 \\
0 & 0 & 0 & 1
\end{pmatrix}
\begin{pmatrix} ct \\ x \\ y \\ z \end{pmatrix} .
\tag{36.4}
$$

Throughout this chapter, we will simplify our notation from β_v and γ_v to β and γ as confusion is unlikely to occur. From (36.4) one finds immediately

$$
\frac{c\partial t'}{c\partial t} = \frac{\partial x'}{\partial x} = \gamma, \qquad \frac{c\partial t'}{\partial x} = \frac{\partial x'}{c\partial t} = -\gamma\beta, \qquad \frac{\partial y'}{\partial y} = \frac{\partial z'}{\partial z} = 1.
$$

All other derivatives are zero. Thus,

$$
\begin{pmatrix} \dfrac{\partial}{c\partial t} \\[2ex] \dfrac{\partial}{\partial x} \\[2ex] \dfrac{\partial}{\partial y} \\[2ex] \dfrac{\partial}{\partial z} \end{pmatrix}
=
\begin{pmatrix}
\gamma & -\gamma\beta & 0 & 0 \\
-\gamma\beta & \gamma & 0 & 0 \\
0 & 0 & 1 & 0 \\
0 & 0 & 0 & 1
\end{pmatrix}
\begin{pmatrix} \dfrac{\partial}{c\partial t'} \\[2ex] \dfrac{\partial}{\partial x'} \\[2ex] \dfrac{\partial}{\partial y'} \\[2ex] \dfrac{\partial}{\partial z'} \end{pmatrix} .
\tag{36.5}
$$

Maxwell's equations read

$$\nabla \cdot \mathbf{E} = \frac{1}{\varepsilon_0} \rho \qquad (36.6a)$$

$$\nabla \times \mathbf{E} = -\frac{\partial \mathbf{B}}{\partial t} \qquad (36.6b)$$

$$\nabla \cdot \mathbf{B} = 0 \qquad (36.6c)$$

$$\nabla \times \mathbf{B} = \mu_0 \left(\mathbf{j} + \epsilon_0 \frac{\partial \mathbf{E}}{\partial t} \right). \qquad (36.6d)$$

ε_0 and μ_0 are the dielectric and magnetic constants, respectively. Both are Lorentz invariant, like c. We begin with the x-component of (36.6b),

$$\frac{\partial E_z}{\partial y} - \frac{\partial E_y}{\partial z} = -\frac{c \partial B_x}{c \partial t},$$

apply (36.5), and rearrange,

$$\gamma \frac{c \partial B_x}{c \partial t'} - \gamma \beta \frac{c \partial B_x}{\partial x'} + \frac{\partial E_z}{\partial y'} - \frac{\partial E_y}{\partial z'} = 0. \qquad (36.7)$$

Then we take (36.6c), multiplied by c,

$$\frac{c \partial B_x}{\partial x} + \frac{c \partial B_y}{\partial y} + \frac{c \partial B_z}{\partial z} = 0,$$

and transform it with (36.5),

$$-\gamma \beta \frac{c \partial B_x}{c \partial t'} + \gamma \frac{c \partial B_x}{\partial x'} + \frac{c \partial B_y}{\partial y'} + \frac{c \partial B_z}{\partial z'} = 0. \qquad (36.8)$$

With these transformations only the coordinate differentials have been transformed, not the field differentials. Now we add β times (36.8) to (36.7) to obtain

$$\gamma (1 - \beta^2) \frac{c \partial B_x}{c \partial t'} + \frac{\partial}{\partial y'} (E_z + \beta c B_y) - \frac{\partial}{\partial z'} (E_y - \beta c B_z) = 0$$

$$\frac{\partial}{\partial y'} [\gamma (E_z + v B_y)] - \frac{\partial}{\partial z'} [\gamma (E_y - v B_z)] = -\frac{\partial B_x}{\partial t'}.$$

Comparison with (34.10) gives

$$\frac{\partial E'_z}{\partial y'} - \frac{\partial E'_y}{\partial z'} = -\frac{\partial B'_x}{\partial t'}.$$

This is the x-component of (36.6b) in the S'-frame, Q.E.D. Now we add β times (36.7) to (36.8) and compare with (34.10),

$$\gamma(1 - \beta^2)\frac{c\partial B_x}{\partial x'} + \frac{\partial}{\partial y'}(cB_y + \beta E_z) + \frac{\partial}{\partial z'}(cB_z - \beta E_y) = 0$$

$$\frac{\partial B_x}{\partial x'} + \frac{\partial}{\partial y'}\left[\gamma\left(B_y + \frac{v}{c^2}E_z\right)\right] + \frac{\partial}{\partial z'}\left[\gamma\left(B_z - \frac{v}{c^2}E_y\right)\right] = 0$$

$$\frac{\partial B'_x}{\partial x'} + \frac{\partial B'_y}{\partial y'} + \frac{\partial B'_z}{\partial z'} = 0.$$

This is (36.6c) in the S'-frame, Q.E.D. We continue by taking the x-component of (36.6d),

$$\frac{\partial B_z}{\partial y} - \frac{\partial B_y}{\partial z} = \mu_0\left(j_x + \varepsilon_0\frac{c\partial E_x}{c\partial t}\right),$$

transform it with (36.5), and rearrange

$$-\mu_0\varepsilon_0\gamma\frac{c\partial E_x}{c\partial t'} + \mu_0\varepsilon_0\gamma\beta\frac{c\partial E_x}{\partial x'} + \frac{\partial B_z}{\partial y'} - \frac{\partial B_y}{\partial z'} = \mu_0\gamma\beta\rho'c + \mu_0\gamma j'_x.$$

For the $\mu_0 j_x$-term, the inverse form of (32.4), namely

$$\begin{pmatrix} \rho c \\ j_x \\ j_y \\ j_z \end{pmatrix} = \begin{pmatrix} \gamma & \gamma\beta & 0 & 0 \\ \gamma\beta & \gamma & 0 & 0 \\ 0 & 0 & 1 & 0 \\ 0 & 0 & 0 & 1 \end{pmatrix} \begin{pmatrix} \rho'c \\ j'_x \\ j'_y \\ j'_z \end{pmatrix},$$

was used. With $\dfrac{1}{c^2} = \mu_0\varepsilon_0$ follows

$$-\gamma\frac{c\partial E_x}{c\partial t'} + \gamma\beta\frac{c\partial E_x}{\partial x'} + \frac{c^2\partial B_z}{\partial y'} - \frac{c^2\partial B_y}{\partial z'} = \frac{1}{\varepsilon_0}\gamma\beta\rho'c + \frac{1}{\varepsilon_0}\gamma j'_x. \quad (36.9)$$

Finally, we take (36.6a), multiplied by c,

$$\frac{c\partial E_x}{\partial x} + \frac{c\partial E_y}{\partial y} + \frac{c\partial E_z}{\partial z} = \frac{1}{\varepsilon_0}\rho c,$$

and transform it,

$$-\gamma\beta\frac{c\partial E_x}{c\partial t'} + \gamma\frac{c\partial E_x}{\partial x'} + \frac{c\partial E_y}{\partial y'} + \frac{c\partial E_z}{\partial z'} = \frac{1}{\varepsilon_0}\gamma\rho'c + \frac{1}{\varepsilon_0}\gamma\beta j_x'. \quad (36.10)$$

Then we take (36.9), subtract β times (36.10), and compare with (34.10),

$$-\gamma(1-\beta^2)\frac{c\partial E_x}{c\partial t'} + \frac{\partial}{\partial y'}(c^2 B_z - \beta c E_y) - \frac{\partial}{\partial z'}(c^2 B_y + \beta c E_z) = \frac{1}{\varepsilon_0}\gamma(1-\beta^2)j_x'$$

$$\frac{\partial}{\partial y'}\left[\gamma\left(B_z - \frac{v}{c^2}E_y\right)\right] - \frac{\partial}{\partial z'}\left[\gamma\left(B_y + \frac{v}{c^2}E_z\right)\right] = \frac{1}{c^2}\frac{\partial E_x}{\partial t'} + \mu_0 j_x'$$

$$\frac{\partial B_z'}{\partial y'} - \frac{\partial B_y'}{\partial z'} = \mu_0\left(j_x' + \varepsilon_0\frac{\partial E_x'}{\partial t'}\right).$$

This is the x-component of (36.6d) in the S'-frame, Q.E.D. Now we take (36.10), subtract β times (36.9), and compare with (34.10),

$$\gamma(1-\beta^2)\frac{c\partial E_x}{\partial x'} + \frac{\partial}{\partial y'}(cE_y - \beta c^2 B_z) + \frac{\partial}{\partial z'}(cE_z + \beta c^2 B_y) = \frac{1}{\varepsilon_0}\gamma(1-\beta^2)\rho'c$$

$$\frac{\partial E_x}{\partial x'} + \frac{\partial}{\partial y'}[\gamma(E_y - vB_z)] + \frac{\partial}{\partial z'}[\gamma(E_z + vB_y)] = \frac{1}{\varepsilon_0}\rho'$$

$$\frac{\partial E_x'}{\partial x'} + \frac{\partial E_y'}{\partial y'} + \frac{\partial E_z'}{\partial z'} = \frac{1}{\varepsilon_0}\rho'.$$

This is (36.6a) in the S'-frame, Q.E.D.

Problems for Part IV

1. **Distance in space-time**
   Given are events $E_1(t = 9\mathrm{a} \mid x = 3\,\mathrm{ly})$ and $E_2(t = 3\mathrm{a} \mid x = 6\,\mathrm{ly})$.
   S' is moving at a velocity of $\beta = 0.8$ relative to S.

   (a) Calculate E_1' and E_2', the Lorentz transformations of E_1 and E_2.
   (b) Calculate the distances $E_1 E_2$ and $E_1' E_2'$ with the four-dimensional scalar product of the time-position vector with itself.

2. **Completely inelastic collision of two particles**
   Calculate for example 1 in Chapter 29 the value of m_c for $m_a = m_b = 1$
   and $\beta_a; \beta_b = \pm 0.90; \pm 0.50; \pm 0.10$, i.e. for 36 pairs of values.

3. **Pair annihilation**
   Calculate $E_c/m_0 c^2$ and $E_d/m_0 c^2$ for $E_a/m_0 c^2 = 1; 2; 3; 4; 5; 6$. Toward
   which asymptotic values do these functions tend? Draw the functions.

4. **Compton effect**
   Calculate example 8 in Chapter 29 again, on the assumption that the
   electron is not at rest but has the velocity u_b.

5. **Four-velocity and four-acceleration**

   (a) By multiplying (30.2) and (30.12) component by component,
   show that $U^\mu A_\mu = 0$ holds. Also give a fast proof of this by differentiating (30.6) with respect to proper time τ.

(b) Insert arbitrary numbers into $U^\mu A_\mu = 0$ to show that all 6 variables, $u_x, u_y, u_z, a_x, a_y, a_z$, are independent, i.e. 6 degrees of freedom exist.

(c) Show that in 2-dimensional (3-dimensional) scalar products of the form $\mathbf{a} \cdot \mathbf{b} = 0$, $\mathbf{a}, \mathbf{b} \in \Re$, only 3 (5) variables can be chosen independently.

6. **Four-acceleration**

By analogy to (30.3), make a Lorentz transformation of the four-acceleration for the special case $\mathbf{u} = (u_x, 0, 0)$ in order to obtain the transformation formulas for acceleration, as given by (11.2). Hint: Use (9.1) and (25.3).

7. **Circular accelerator**

Protons with a relativistic energy of 300 GeV are moving through a circular accelerator of 1000 m radius.

(a) Calculate their velocity and frequency.

(b) Calculate their centripetal acceleration and their centripetal proper acceleration.

(c) As a comparison, calculate the acceleration of free fall on a neutron star of mass 2.8×10^{30} kg and radius 10 km.

8. **Rotational invariance and Lorentz invariance**

Show that the phase φ of a plane wave is (a) rotationally invariant, and (b) Lorentz invariant.

9. **Charge density-current density vector**

(a) Give a well-founded estimate of the charge density ρ and the electric current density $\mathbf{j}$ in a current-carrying filament at your home. Then calculate the velocity of the conduction electrons.

(b) Write down the charge density-current density vector and transform it into a reference frame which moves as fast as the conduction electrons. By what fraction does the positive charge density change? Then, read Chapter 5 again.

10. **Lorentz transformation of forces and accelerations**

    In the S-frame, an electron of rest mass m_0 is moving at a velocity of $u = (0, u_y, 0)$. According to (22.6), the force $f = (f_x, 0, 0)$ exerts an acceleration of $a = (f_x/\gamma_u m_0, 0, 0)$ on the electron. The S'-frame is moving at $v = (v, 0, 0)$ relative to the S-frame. Use (33.6) to calculate the force and (11.2) to calculate the acceleration in the S'-frame. Does $f' = m'a'$ hold?

11. **Lorentz transformation of forces and fields**

    In the S-frame, an electron is moving at a velocity of $u = (u, 0, 0)$ through a combination of electric field $E = (0, 0, E_z)$ and magnetic field $B = (0, B_y, 0)$. The S'-frame is moving at $v = (v, 0, 0)$ relative to the S-frame.

    (a) Show that the electron's motion is not deflected if $u = -E_z/B_y$, because the Coulomb force, f_C, and the Lorentz force, f_L, are of equal but opposite size.

    (b) Use (34.10) to transform both fields into the S'-frame, take (9.1), and show that $u' = -E'_z/B'_y$ turns to $u = -E_z/B_y$.

12. **Lorentz transformation of the electric and the magnetic field**

    Show that $E' \cdot B' = E \cdot B$ is also valid

    (a) in the non-relativistic approach, (34.19),

    (b) in the most general relativistic case, (34.18), i.e. for $v = (v_x, v_y, v_z)$.

Solutions

Part I

1. (1.1): 1.6×10^{-15}; (1.2): 4.0×10^{-16}
2. (a) $v = 34.3\,\text{km/s}$; $T = 212$ d; (b) $d = 1.27 \times 10^3$ ly
4. $\gamma = 1 + 4.3 \times 10^{-15}$; $\gamma = 1 + 4.3 \times 10^{-13}$; $\gamma = 1 + 3.6 \times 10^{-10}$; $\gamma = 1.00000035$; $\gamma = 1.000050$; $\gamma = 1.0050$; $\gamma = 2.3$; $\gamma = 7.1$; $\gamma = (2\varepsilon)^{-1/2}$
5. $\tau = 9.4$ s; $\tau = 0$; no proper time defined as events have a space-like separation
6. $\beta = 0.97$; $\beta = 1 - 6.9 \times 10^{-10}$; $\beta = 1 - 8.7 \times 10^{-14}$
7. $t - \tau = 6.2 \times 10^{-9}$ s; $t - \tau = 6.2 \times 10^{-8}$ s; $t - \tau = 1.8 \times 10^{-6}$ s
8. $\tau_B - \tau_A = 196 \times 10^{-9}$ s; $\tau_B - \tau_A = 115 \times 10^{-9}$ s
9. $v = 144\,\text{km/h}$; $v = 163\,\text{km/h}$
10. $t = \sqrt{\dfrac{1-\beta}{1+\beta}}\dfrac{\ell'}{c}$; $t' = \dfrac{\ell'}{c}$
11. (b) $A = 12\,\text{cm}^2$
12. (a) $V = (a^2 + b^2)^2$; (b) $V' = (a^2 + b^2)^2/\gamma$
14. $N = 474/\text{s}$
15. (a) $v = 9.5 \times 10^{-5}\,\text{m/s}$; $\gamma = 1 + 5.0 \times 10^{-26}$;
    $\rho'_\pm = \pm(1 \pm 5.0 \times 10^{-26}) \times 1.34 \times 10^{10}\,\text{C/m}^3$;
    $\rho' = \rho'_+ + \rho'_- = 1.34 \times 10^{-15}\,\text{C/m}^3$
    (b) $f_L = 3.03 \times 10^{-28}$ N; $f'_C = 3.03 \times 10^{-28} \times (1 + 5.0 \times 10^{-26})$ N
17. (a) $E'_1(5.5| - 6.5)$; $E'_2(-0.5|3.5)$; $E'_3(0.5| - 3.5)$; $E'_4(-5.5|6.5)$
18. (a) $E''_1(-0.5|-3.5)$; $E''_2(5.5|6.5)$; $E''_3(-5.5|-6.5)$; $E''_4(0.5|3.5)$
19. $S \rightarrow S'_1$: Gal.(2.00|4.98), Lor.(1.95|4.98);
    $S \rightarrow S'_2$: Gal.(2.00|4.80), Lor.(1.51|4.82);
    $S \rightarrow S'_3$: Gal.(2.00|3.20), Lor.(−5.74|7.34)

20. E_1 and E_2: $\beta_2 = 0.43$; E_1 and E_3: no reference frame as events have time-like separation

22. $d = 1.5 \times 10^{12}$ km $= 58$ light-days $= 0.16$ ly

23. S-frame: $(2.5\,\text{s}|0.5\,\text{ls})$ and $(3.9\,\text{s}|0.8\,\text{ls})$;
    S'-frame: $(2.4\,\text{s}|0.0\,\text{ls})$ and $(3.8\,\text{s}|0.0\,\text{ls})$

24. $t = 40$ min

Part II

1. (a) $u_x = 0.89\,c$; (b) $u_y = -0.13\,c$; $u_z = 0.19\,c$; $u = 0.92\,c$

3. (b) 1.5×10^{12} km $= 57$ light-days $= 0.16$ ly

4. $\left(\dfrac{d\theta'}{d\theta}\right)^2_{\theta=0°} = \dfrac{1+\beta}{1-\beta} = 19;\ 199;\ 1999;$

   $\left(\dfrac{d\theta'}{d\theta}\right)^2_{\theta=180°} = \dfrac{1-\beta}{1+\beta} = \dfrac{1}{19};\ \dfrac{1}{199};\ \dfrac{1}{1999}$

5. $a_x = 2.64\,\text{m/s}^2$; $a_y = 4.42\,\text{m/s}^2$; $a_z = 3.66\,\text{m/s}^2$

7. (a) $x' = \ell \cosh\left(\dfrac{\alpha\tau}{c}\right) + \dfrac{c^2}{\alpha}\left(\sqrt{1 + \dfrac{\alpha^2}{c^4}\sinh^2\left(\dfrac{\alpha\tau}{c}\right)\ell^2} - 1\right)$

   (b) $\dfrac{dx'}{d\tau} = \dfrac{\ell\alpha}{c}\sinh\left(\dfrac{\alpha\tau}{c}\right) + \dfrac{\dfrac{\ell^2\alpha^2}{c^3}\sinh\left(\dfrac{\alpha\tau}{c}\right)\cosh\left(\dfrac{\alpha\tau}{c}\right)}{\sqrt{1 + \dfrac{\ell^2\alpha^2}{c^4}\sinh^2\left(\dfrac{\alpha\tau}{c}\right)}},$

   (c) $x' \to \infty$, $\dfrac{dx'}{d\tau} \to \infty$

8. (14.3): 1.73 s and 0.58 s; (2.1): 1.15 s

9. $t = 0.71$a; $\tau = 0.66$a

11. e.g. $t = -3$ s: actual position -2.4 ls; apparent position -14.2 ls; stretch factor 1.02; $v^*/c = 1.36$
    e.g. $t = +6$ s: actual position $+4.8$ ls; apparent position $+1.4$ ls; stretch factor 0.37; $v^*/c = 0.50$

Part III

1. $\beta = 0.995$; $\beta = 0.99995$; $\beta = 0.9999995$
2. e.g. $\tau = 2a$: $x = 3.0 \,\text{ly}$; $\beta = 0.97$; $a = 0.14 \,\text{m/s}^2$; $M_f/M_i = 0.12$; $M_f/M_i = 0.51$; $m/m_0 = 4.2$
3. e.g. $u/c = 0.10$: (a) $a_x = 0.985 \,\text{m/s}^2$; $a_y = a_z = 0$;
   (b) $a_y = 0.995 \,\text{m/s}^2$; $a_x = a_z = 0$
4. e.g. $r = 10^{-12} \,\text{m}$: $\beta = 0.075$
5. (a) $E_{\text{kin}} = \dfrac{\pi}{4}\omega^2 \rho D R^4$;

   (b) $E_{\text{kin}} = 2\pi c^4 \dfrac{1}{\omega^2}\rho D \left(1 - \sqrt{1 - \dfrac{\omega^2 R^2}{c^2}} - \dfrac{\omega^2 R^2}{2c^2}\right)$;

   $E_{\text{kin}}^{\max} = \pi c^2 \rho D R^2$

7. $E_c = 324 \,\text{keV}$; $E_d = 545 \,\text{keV}$; $p_c = -324 \,\text{keV}/c$; $p_d = 189 \,\text{keV}/c$
8. $\Delta E = 0.99 \,\text{GeV}$; $E_{\text{kin}} = 8.2 \,\text{MeV}$
9. $E_{\text{kin}} = 0.050 \,\text{eV}$; $u = 300 \,\text{m/s}$
10. (a) $E = 2m_0 c^2$; $\mathbf{p} = (m_0 c, m_0 c, m_0 c)$;
    (b) $E' = \sqrt{3}m_0 c^2$; $\mathbf{p'} = (0, m_0 c, m_0 c)$;
    (c) $E'' = m_0 c^2$; $\mathbf{p''} = (0, 0, 0)$

Part IV

1. (a) $E_1'(11a|-7\,\text{ly})$; $E_2'(-3a|6\,\text{ly})$; (b) $3\sqrt{3}\,\text{ly}$
2. e.g. $\beta_a = 0.90$ and $\beta_b = 0.50$: $m_c = 2.22$
3. e.g. $\dfrac{E_a}{m_0 c^2} = 3$: $\dfrac{E_c}{m_0 c^2} = 3.41$; $\dfrac{E_d}{m_0 c^2} = 0.59$;

   $\dfrac{E_c}{m_0 c^2} \to \dfrac{E_a}{m_0 c^2} + \dfrac{1}{2}$, $\dfrac{E_d}{m_0 c^2} \to \dfrac{1}{2}$

4. $(1 - \beta_b)\lambda' - (1 - \beta_b \cos\theta)\lambda = \dfrac{h}{\gamma_b m_0 c}(1 - \cos\theta)$
7. (a) $\beta = 0.9999951$; $f = 47.7 \,\text{kHz}$;
   (b) $a = 9.00 \times 10^{13} \,\text{m/s}^2$; $\alpha = 9.21 \times 10^{18} \,\text{m/s}^2$;
   (c) $1.9 \times 10^{12} \,\text{m/s}^2$
9. (a) $\rho_\pm \approx \pm 10^{10} \,\text{C/m}^3$; $j_x \approx 10^6 \,\text{A/m}^2$; $v \approx -10^{-4} \,\text{m/s}$;
   (b) $\rho'/\rho_+ \approx 10^{-25}$
10. $\mathbf{f'} = (f_x, 0, 0)$; $\mathbf{a'} = \left(\dfrac{f_x}{\gamma_u \gamma_v^3 m_0}, \dfrac{\beta_u \beta_v f_x}{\gamma_u \gamma_v^2 m_0}, 0\right)$; no

Further Reading

G. Barton (1999) *Introduction to the Relativity Principle*, Wiley.

L. Bergmann and C. Schaefer (2001) *Optics: Of Waves and Particles*, CRC Press.

D. Bohm (2006) *The Special Theory of Relativity*, Routledge.

H. Bondi (1980) *Relativity and Common Sense*, Dover Publications.

A. Einstein (2003) *The Meaning of Relativity*, Routledge.

A. Einstein (2007) *Relativity: The Special and the General Theory*, Dodo Press.

M. Fayngold (2002) *Special Relativity and Motions Faster than Light*, Wiley-VCH.

R. P. Feynman, R. B. Leighton, and M. Sands (2005) *The Feynman Lectures on Physics*, 3 volumes, Benjamin Cummings.

A. P. French (1968) *Special Relativity*, CRC Press.

D. Giulini (2005) *Special Relativity: A First Encounter: 100 years since Einstein*, Oxford University Press.

W. Greiner (2004) *Classical Mechanics: Point Particles and Relativity*, Springer.

R. d'Inverno (1992) *Introducing Einstein's Relativity*, Clarendon Press.

J. D. Jackson (1998) *Classical Electrodynamics*, Wiley.

J. B. Kogut (2001) *Introduction to Relativity: For Physicists and Astronomers*, Academic Press.

L. D. Landau and E. M. Lifshitz (1995) *The Classical Theory of Fields, (Course of Theoretical Physics, Volume 2)*, Butterworth–Heinemann.

N. D. Mermin (2005) *It's About Time: Understanding Einstein's Relativity*, University Presses of California.

C. Møller (1952) *The Theory of Relativity*, Oxford University Press.

Th. A. Moore (1995) *A Traveler's Guide to Spacetime*, McGraw-Hill.

R. A. Mould (2001) *Basic Relativity*, Springer.

R. Resnick (1968) *Introduction to Special Relativity*, Wiley.

W. Rindler (1991) *Introduction to Special Relativity*, Clarendon Press.

W. Rindler (2006) *Relativity: Special, General, and Cosmological*, Oxford University Press.

F. N. H. Robinson (1996) *An Introduction to Special Relativity and Its Applications*, World Scientific.

U. E. Schröder (1990) *Special Relativity*, World Scientific.

P. M. Schwarz, J. H. Schwarz (2004) *Special Relativity: From Einstein to Strings*, Cambridge University Press.

E. F. Taylor and J. A. Wheeler (1992) *Spacetime Physics*, Palgrave Macmillan.

P. A. Tipler and R. A. Llewellyn (2003) *Modern Physics*, Palgrave Macmillan.

W. S. C. Williams (2002) *Introducing Special Relativity*, Taylor & Francis.

N. M. J. Woodhouse (2002) *Special Relativity*, Springer.

Index